RANCHES OF COLORADO

PROJECT SPONSORS

FirstBank | Xcel Energy | 9NEWS KUSA

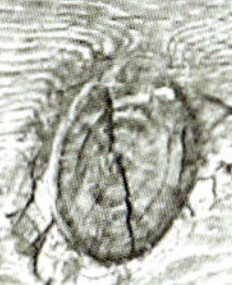

5280 Denver's Magazine | Mirr Ranch Group, LLC | Johnson Storage and Moving Company

IN MEMORY OF JAMES B. MEADOW

The Colorado Land Trust Community

From the formation of the Colorado Stock Growers Association in 1867; to John Wesley Iliff, John W. Prowers, and the cattle barons of the 1870s; to the inaugural National Western Stock Show in 1905, ranching has long been the heritage of Colorado. Similarly, from being the birthplace of the U.S. Wilderness system at Trapper's Lake in 1920, to supporting the Great Outdoors Colorado Trust Fund, to creating an innovative conservation easement program, Colorado has proved a deep and aggressive commitment to conservation. This book illustrates the preservation of both these legacies; most of the ranchers in this book have committed to protecting their land through conservation easements. We hope *Ranches of Colorado* will serve as a tribute to the accomplishments of both ranching and conservation, and a stimulant to their continued partnership. A portion of the proceeds of this book will go to three organizations, detailed below, that have played a key role in its creation, as well as in the preservation of the unique Colorado places it celebrates.

Colorado Cattlemen's Agricultural Land Trust

The Colorado Cattlemen's Agricultural Land Trust (CCALT) was formed in 1995 by the membership of the Colorado Cattlemen's Association, to help ranchers and farmers protect their agricultural lands for future generations. It is a land trust made up of landowners, by landowners, and for landowners. CCALT is proud to have so many of the ranchers we have worked with featured in *Ranches of Colorado*.

CCALT is often a bridge between the conservation and agricultural communities, and we play a key role in protecting Colorado's working landscapes. We exist to help landowners preserve not only their lands, but their heritage—the lifeblood of our rural communities—as well. CCALT's success would not be possible without the ranching families of Colorado and their tireless devotion to the land. With our support, increasing numbers of farmers and ranchers are embracing the conservation easement tool.

CCALT is the only organization dedicated exclusively to protecting Colorado's agricultural lands—which make up 85 percent of the state's private land. To date, CCALT has partnered with more than 200 ranchers and farmers and protected more than 325,000 productive acres. We are very encouraged by the shift in attitude from environmental groups and everyday citizens, who increasingly realize the benefits of agriculture to Colorado's landscapes. CCALT takes great pride in the fact that we were recently accredited by the Land Trust Accreditation Commission, demonstrating the organization's commitment to excellence in land conservation and governance.

Despite our success, Colorado continues to lose agricultural land at an alarming rate. In fact, Colorado ranks third in the nation for conversion of agricultural land, and over the past decade, more than 1.8 million acres (548 acres a day) were converted to other uses. Fortunately, there is reason to be optimistic as we look forward. We are encouraged by several new conservation initiatives that will protect more land—from the lush hay meadows and valleys of northwest Colorado to the vast shortgrass prairies of the eastern plains. CCALT focuses our efforts on landscapes with long-term agricultural viability to ensure a future for Colorado's ranches and farms and an improved quality of life for all of our citizens.

Three areas featured in *Ranches of Colorado* exemplify CCALT efforts to protect Colorado's working landscapes:

SAGUACHE CREEK: Since 1998, more than a dozen landowners have partnered with CCALT in the protection of fifteen properties along Saguache Creek on the northwest side of the San Luis Valley. This valley represents the longest stretch of undeveloped highway in western Colorado.

ELK RIVER VALLEY: In the early 1990s, a group of ranchers in this valley north of Steamboat Springs banded together to protect their ranches with conservation easements, representing one of the first efforts of its kind in Colorado. This idea led to the formation of CCALT. Today CCALT protects more than 6,000 acres in the Elk River Valley.

SOUTHEAST COLORADO: The shortgrass prairie plains of southeastern Colorado give way to spectacular red rock canyons and volcanic mesas to the east of Trinidad. CCALT was one of the first conservation groups to work in this impressive landscape where more than 140,000 acres are now protected.

Colorado Cattlemen's Agricultural Land Trust
8833 Ralston Road
Arvada, CO 80002
www.ccalt.org

Colorado Open Lands

Colorado Open Lands' mission is to preserve the significant agricultural and open lands and natural heritage of Colorado through private and public partnerships, innovative land conservation techniques, and strategic leadership. Our mission has been instrumental in guiding us to protect western lands for wildlife, agriculture, and other uses.

Land conservation always begins with the landowner. Across Colorado, our organization has been fortunate to work with hundreds of landowners committed to the conservation of their land. Their commitment may originate from a desire to preserve generations of family heritage. It may stem from a love of the land itself. It may reflect their dedication to perpetuating agriculture. It may come from the ranchers' hopes to raise children in their lifestyle and the lifestyle of their parents. More than likely, it's a result of all these factors.

Often, these individuals and families are third and fourth-generation ranchers and farmers who raise cattle and crops on both dry and irrigated land. Many are over 55 years old and often have debt or health concerns. Mounting development pressures are challenging these agricultural producers' ability to pay taxes on increasingly valuable land, not to mention making livestock management difficult in increasingly residential areas.

Founded in 1981, Colorado Open Lands works cooperatively in all types of creative private and public collaborations. Our organization has been particularly effective in developing long-term, multiple-partner projects to conserve large landscapes of statewide and national significance. As a leader in statewide and national land conservation initiatives, our problem-solving, non-adversarial approach to land preservation has been instrumental to our success.

We also value accountability. Consequently, 98 percent of our revenue goes directly to our land protection programs. In addition, Colorado Open Lands is among the first group of land trusts to be accredited by the Land Trust Accreditation Commission. Accreditation provides public recognition of standards for governance and organizational management that typify best practices for land trusts.

Colorado Open Lands works to protect active family ranches from development, thereby preserving the land, as well as enabling the long-term continuation of each property's agricultural use. We focus our efforts in specific geographic areas such that, over the long run, multiple family-owned farms and ranches will be protected in one community and further increase the likelihood that the agricultural community as a whole will remain viable. The Gunnison Valley, South Platte, and Pueblo County ranches featured in this book are all examples of Colorado Open Lands' community conservation efforts.

With ranching and farming woven throughout Colorado's history as well as its present culture, preservation of these agricultural lands is critical to preserving our economy, our quality of life, our collective culture, and the historic appearance and scenic quality of our communities.

"We see this project as adding to the sustainability of our cattle business and our children's ability to inherit the ranch. It is also another example of diverse groups working together to achieve a common goal of conservation."—Steve Wooten, Beatty Canyon Ranch

Colorado Open Lands
274 Union Boulevard, Suite 320
Lakewood, CO 80228
www.coloradoopenlands.org

Colorado Coalition of Land Trusts

The Colorado Coalition of Land Trusts (CCLT) was established in 1991 by concerned citizens and land trust practitioners in order to create a collective voice for all of Colorado's conservation groups. The organization is governed by an active eleven-member board consisting of representatives from member organizations across the state, as well as other conservation professionals and individuals who care passionately about our mission. As the collective voice for conservation in Colorado, we support the efforts of our fifty-four member land trusts and local government open space programs as they work to preserve Colorado's wildlife, working farms and ranches, and significant natural landscapes.

The need for land conservation has skyrocketed in recent years, as have the number of land trusts and active local-government open-space programs. Colorado currently has an estimated 4.8 million residents, a number that is expected to increase by 1 million in the next ten years. As the population escalates, the demand for parks and open space grows. CCLT works to secure public policy support for increasing open space and provides on-going education for members to ensure that conservation organizations have the knowledge and skills to preserve land in a way that is cost-effective and lasting. By joining together, we advance land conservation, and to date have ensured that over 2 million acres of the state's spectacular lands have been preserved, an area nearly seven times the size of Rocky Mountain National Park!

Through CCLT's professional development workshops and conferences, both the number of land trusts in Colorado and the capacity of those land trusts has increased dramatically—according to a census by the Land Trust Alliance, land donations in Colorado have tripled since 1999. In recent years, we have published guides on phasing conservation easement transactions, appraising conservation easements, and funding opportunities for open space projects. In 2008 we published *Mineral Development and Land Conservation, A Handbook for Conservation Professionals*. We hold annual conferences and training sessions with experts in the field, on issues such as stewardship of preserved lands and best practices for new land conservation deals.

On the public policy front, CCLT spearheaded the passage of one of the state's most important land preservation tools, the Colorado conservation easement income tax credit. We have successfully guided four additional initiatives through the Colorado General Assembly that have expanded the program and increased its accountability and transparency. Most notable in recent history is House Bill 08-1353, which increases accountability by land trusts and conservation easement appraisers and requires that the Colorado Division of Real Estate review all conservation easement appraisals.

Colorado Coalition of Land Trusts
1245 East Colfax Avenue, Suite 203
Denver, CO 80218
www.cclt.org

[A list of CCLT member organizations appears in the back of this book]

Like many environmentalists, I questioned whether the eating of beef was worth the bad things that cattle could do to the land.

The science of rangeland ecology has advanced greatly in recent years, not only in the West but around the world, and it draws its substance in large part from the history of wild animals that have grazed on Earth for thousands of years. Ecologists study the relationship between living things and their environment, and rangeland ecologists study, among other things, how herbivores such as mammoths, mega-beavers, buffalo, and elk, existed or exist by eating plants. Conversely, they study how these creatures affect the land. Grazing, defecating, urinating, and sharp hooves disturbing solid ground are all part of the process that keeps rangelands healthy and reproductive. We know that landscapes from which herbivores have been removed are not reproductive, and are less biologically diverse. The elimination of buffalo from the Great Plains in the 19th century is a prime example of this.

Like a rangeland ecologist, I, photographer and observer, began to look at the way that the land of the West is currently used. Most western ranches operate this way: The family owns "bottomland" along a creek or river where ditches are used to get the water out of the river and into the meadows to grow hay. The hay is used to feed the cattle in the winter when the grasses are dormant. The cattle remain at these lower elevations in the winter because there's less snow and the rancher is close-by in case assistance is needed when cows are giving birth to calves in the frigid months of February, March, and April. Because hay grows on the rancher's property in the summer, the cattle must be taken some place else to fatten up — public lands. Cows are sent to the high country in the summer, usually to national forest or BLM land, and the rancher pays the government for the right to do this. These "leases" are renewed regularly, and a rancher has as much title to them as he does to the bottomland he owns in fee simple.

Without a place to graze the cows in the summer, a ranch cannot exist. The bucolic river valleys that we love to view along our western highways, with their rickety fences and barns, their green meadows, and summer hay bales, would most likely end up in other uses, namely for homes and ranchettes, if there was no public lands grazing. This would not be good. Not good for the people and families who enjoy a life outdoors tending the land and their livestock. Not good for tourism and the people from all over the world who drive the highways and byways of the American West spending money along the way. And certainly not good for biodiversity. Elk, deer, and other migratory creatures depend upon private ranch land for winter habitat. If those ranches were condos, there would be no place for them to go.

It is ironic that I began my Colorado life on a ranch in the summer of 1967, then disowned ranching for a while because I thought it was destructive, and then returned to ranching in order to promote it by undertaking this *Ranches of Colorado* project. I chose to publish this book in order to show people as graphically as I can what's at stake. To do this, I spent two years photographing fifty working Colorado ranches that exist in almost every type of landscape known in Colorado. In this book are ranches on the eastern Great Plains, ranches in the foothills of the Rocky Mountains, ranches in almost every major mountain valley, and ranches in and above the river canyons that define the western slope of Colorado. Most are owned by "land rich, cash poor" families who have been ranching and farming for several generations — in many cases since the last half of the 19th century. The rest are owned by families who do not live at the ranch, but instead hire ranch managers to operate them as working ranches. Both are in the business to raise cattle and promote the beef industry, and most have protected their land forever from development with conservation easements.

A conservation easement is a restriction on the deed of ownership in which the landowner forfeits his or her right to develop or subdivide the land. Such a restriction usually exists in perpetuity and precludes development of the property by succeeding generations, even by other owners. Typically, such protection lowers the value of a ranch. However, in places such as Colorado's Roaring Fork Valley, it has been shown that demand for "trophy" ranches is high enough to put a premium on intact ranches, and that their value is no less than what the sum of parts of the ranch, if sold, might be. In any case, federal and Colorado state tax laws provide various tax incentives to encourage landowners who own land that has "conservation" value to put it under easement. These laws were made to preserve the lands that define our American heritage.

A landowner who places a conservation easement on their property can deduct the development value — the difference between what a ranch might be worth subdivided and what it is worth intact — from ordinary income taxes, advantageous to people with large incomes. The typical Colorado rancher does not have high income and, therefore, cannot take advantage of this opportunity. Realizing this and wanting to provide more incentive for such families, Colorado enacted a tax credit program in which landowners who place an easement on their property can earn a credit against state income taxes. This credit can be bought and sold in the marketplace, thereby earning cash for the landowner even if his income is not high enough to use the credit personally. In addition, since the value of a landowner's estate typically declines when an easement is created, estate taxes may be lower upon the death of the owner — possibly making it easier for the heirs to retain the ranch rather than have to sell it in order to pay those taxes. But is the state tax credit enough of an incentive for the land rich, cash poor families to want to protect their ranch with an easement? In most cases, probably not, at least not the entire ranch. To bridge this gap, organizations called land trusts pay landowners the development value of the ranch, or a portion thereof, to place a conservation easement on the deed. The landowner "gets their cake, and they get to eat it, too." They now have a portion of the money they would have made selling out to a developer to use for operating costs or as a retirement fund, and they can continue to own and work the ranch. Land trusts exist in most communities in Colorado — there are also national land trusts such as American Farmland Trust. In addition to providing financial incentives, they provide landowners with legal advice necessary to obtaining an easement. The trust is also required to "hold" the easement by monitoring in perpetuity adherence by the landowner to the terms of the easement, such as covenants that govern any allowed development, including the construction of homes on certain sites on the property for family members only.

This book was made possible by the fifty remarkable Colorado ranch families who each allowed me to spend a few days photographing their ranch. I am indebted to them for the opportunity they gave me to tour their own unique piece of Colorado. I met families whose work ethic, personal values of honesty and integrity, passion for their business and industry, and appreciation of their heritage impressed me beyond all of my expectations. I learned that, without exception, each of the families has made a personal commitment to nurture the land that they manage in a sustainable way. Regardless of their respective financial situation, they have not compromised stewardship of the land, in order to allow future generations to operate it productively. All of the families not only possess a remarkable understanding of nature's processes, but also put them to good use without degrading them, in order to maximize production of hay and cattle.

I am also in the debt of James Meadow, writer/reporter for the former *Rocky Mountain News*. I befriended James when my retrospective and repeat photography book, *Colorado 1870-2000*, was published in 1999. The Rocky celebrated the millennium change with a serialization of the book in the paper, and they chose James to write history essays to complement the comparison photographs.

A native New Yorker, James arrived in Denver in 1971 to work for the Rocky. It didn't take long for him to fall in love with the beauty, rhythms, and soul of Colorado. Over the ensuing thirty-eight years he wrote for daily, weekly, and monthly publications, traveling almost everywhere, from the 2002 Winter Olympics in Salt Lake City to an orphanage in Ethiopia. I had always been impressed with his ability to portray the essence of a person by way of an interview — you get to quickly know his subjects as James adroitly filled the spaces between their quotes with his own words, like grout between tiles.

I asked James to interview ten of the fifty ranchers whose ranches are depicted in this book. I felt that if anyone could get these people, who covet their privacy more than most, to open up, it would be James. In 2008 James traveled far and wide to spend a few days living with each of these families. Our goal was to elicit their passion for their ranching heritage, to manifest their personal relationship with the land, and to describe what it's like to be a rancher. In my opinion he succeeded. James is no less than a Picasso with words. His artful use of language makes one feel as if they were standing in the ranch house kitchen sharing a cup of strong coffee with the rancher himself.

My partners in this project are the members of the Colorado land trust community, especially Colorado Cattlemen's Agricultural Land Trust (CCALT) and Colorado Open Lands (COL), which operate statewide; and Colorado Coalition of Land Trusts (CCLT), which represents local land trusts from almost every community in the state. They agreed to find fifty ranches for me to photograph, to introduce me to the families, and to help me promote the project and book. For their participation, I agreed to gain the book publisher's commitment to return a part of the book profits to the land trust community. My partners and I agreed to promote my tour throughout Colorado following book publication to speak publically about saving Colorado's remaining working ranches. I want to thank Chris West from CCALT, Dan Pike from COL, and Jill Ozarski from CCLT, who run their respective organizations, for their participation in the project.

This book does not romanticize the American cowboy and his lonesome lifestyle, nor are the photographs herein made of boots and buckles, and roping and riding. This book is about something more permanent — the land on which the cowboy, the rancher, makes a living by raising cattle. It is about the rancher's regard for the land and its productivity, and his or her admission that no human endeavor can succeed in the long term unless the rules of nature are respected and followed. Through the photographs and words herein, I hope that you will better appreciate the place that ranching resides in Colorado's unique heritage and quality of life, and its influence on our economy and ecology.

[James Meadow died on March 8, 2009, from injuries sustained in a biking accident, three days after completing the final edits of his ten essays in this book.]

Ranches of Colorado *is John Fielder's thirty-ninth book of photographs. He was a progenitor of the Great Outdoors Colorado Trust Fund (GOCO), created in 1992 to redirect Colorado's lottery profits towards protecting the state's natural heritage, including its ranches, and is a recipient of Sierra Club's Ansel Adams Award for Conservation Photography. John Fielder lives in Summit County, Colorado.*

graze on public lands. It is estimated that the 21,000 ranch families that use approximately 30,000 grazing permits on Bureau of Land Management and U.S. Forest Service lands own about 107 million acres of private land. Those permits allow cows and sheep to fatten up on summer grass while hay grows high on the private bottom lands. That hay is what feeds the livestock through the treacherous winter months. Therefore, many ranches cannot operate without the dual use of public and private lands. I am not sure how much the public values ranching, but perhaps they would more so if they knew that public lands grazing keeps private lands out of development, and the West open and resistant to sprawl.

Another societal benefit from this public-private partnership between ranchers and our federal land agencies is the buffering effect of private ranchlands along the public-land borders. For example, in the Southern Rocky Mountains, ranches comprise 21% of all the private lands in the ecoregion. More importantly, however, is that these lands make up 43% of the private lands adjacent to public lands. This observation supports the notion that private ranch lands provide a land-use buffer around our public lands, in essence shielding them from the harmful effects of the rural sprawl that increasingly encircles them.

Robust Cultures

The West is a region of diverse ecosystems, economies, and cultures. Ranching, as a land use and as a culture, has been with us for over 400 years, dating back to the early Spanish colonists who struggled northward over El Paso del Norte and found a home for their livestock near present-day Espanola, New Mexico. If what I have presented in this essay is true, that ranchers and ranching are disproportionately important to the ecology and economy of a West that works, then why are they vilified? Consider this quote by a prominent environmentalist: *"Yes, we are destroying a way of life that goes back one hundred years. But it's a way of life that is one of the most destructive in our country. . . . Ranching is one of the most nihilistic life styles that the planet has ever seen. It should end. Good riddance."*

What does one say to such a final pronouncement of cultural continuation? If ranching is to flourish, persist, or disappear in the West, it should be a conscious decision, based on informed discussions, not due to apathy, neglect, or hate.

Perhaps inflammatory statements such as this are reflections of nothing more than different values. Might some Americans want the public and private lands free of manure, cows, sheep, and fences because they want them for their own uses, such as mountain biking and river rafting? Do some want ranchers and their livestock off the Western ranges because they believe what others have told them, that cows and sheep sandblast land and that cattle barons are arrogant and intolerant of any but their own kind?

What about the Far Right—the New Federalists who are obsessed with spreading their private-property rights hysteria? These radical players in the West throw out incendiary remarks about wildland protection and government land grabs as easily as the Far Left reflexively opposes livestock grazing. Thank goodness for those in the "radical center" who strive to build links across landscapes—connections that run through human and natural communities and cross sociopolitical chasms.

Perhaps the wing nuts at either end of this spectrum stir up dissent because they find it easier to simplify, divide, demean, and demonize. Meanwhile, more and more ranchers are at the core of building new alliances; connecting rural and urban communities while providing food and conserving open space. Whether it be with the Blackfoot Challenge, the Diablo Trust, the Quivira Coalition, the Malpais Borderlands Group, or scores of other watershed-based groups across the West, progressive ranchers are providing leadership in forging a new way of doing business—one that promotes healthy human and natural communities and that offers hope and inspiration rather than conflict and hate.

My sense is that differing values and distorted mythology can obscure facts, and that at the end of the day, emotion may trump judgment. Would it make any difference if the general public was aware that ranchers are stewards of the land, that cows and sheep are tools in the recovery of arid ecosystems, that open space, biodiversity, and county coffers are enriched more from ranching than from the rapidly eclipsing alternatives, ranchettes and recreation? Perhaps.

There are those who say the only difference between ranchers and realtors is a rancher hasn't sold their ranch yet. Are ranchers just developers in sheep's clothing? Certainly there are some that see their ranch as their last cash crop, their 401(k) account. On the other hand, mounting evidence suggests that ranchers care for the West's geography every bit as much as those of us in the cities and suburbs. In Colorado, the state's cattlemen's association has formed a land trust, the Colorado Cattlemen's Agricultural Land Trust, one of the partners of this book and project. To date, 202 conservation easements, totaling over 320,000 acres, have been entrusted to it from ranch families. Indeed, in Colorado, this land trust is second only to The Nature Conservancy in acres protected under conservation easements. Ranching is mostly a sustainable economy; it occupies the dicey ground along the economic margins of profit and loss. So why do ranch families conserve their land, rather than sell it to a developer? Like you and I, they love the wide-open spaces of the West, and feel it's an honorable profession to provide food and ecosystem services to all Americans.

Western ranching has spanned the time scale from the First Americans to the astronauts, avoiding the moving-on mandate of the get-rich-quick industries of mining and logging. Charles Wilkinson, among the most distinguished of our region's scholars, had this to say about the region's ranchers and farmers: *"Yet these industries are the foundation for local economies and provide food for the nation and the world. They preserve open space. As a culture, the people of the ranches and farms have settled in so deeply and for so long that for all practical purposes they are indigenous societies."*

In the heated arguments between ranchers and environmentalists, I will admit to coming down on the rancher's side. In our New West, which is increasingly dominated by urban, suburban, and recently exurban Westerners, it occurs to me that perhaps we could settle this land better than those who conquered the Old West if we listen to the cultures that were here before us (and that endure still). Might we have made a better place of this region if we had slowed down enough to listen to the First Americans? Did they have something to teach us about the region's wildlife, rivers and streams, grasslands and forests? Certainly their ethic that denied any formal ownership of land in favor of being mere "leaseholders" nurtured a sense of sustainability and respect for their descendents. The survival of Native Americans was dependent upon the ecological health of the land and they had to be tuned in to the rhythms of Nature. For example, they used fire on the prairies to drive bison to new grass from season to season in order to create a year-round food supply. In the words of Wendell Berry, *"As important a reason as any to support ranching, farming, irrigating, and logging is that our society will need them as teachers, mentors, and critics in the years to come."*

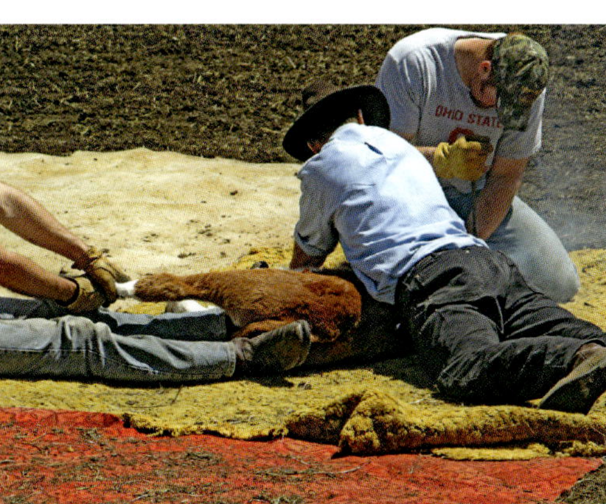

So today, in our haste to remake ourselves once more into the Next West, might we avoid some mistakes by showing respect to the ranching culture? A definitive answer to that question eludes me, but my gut says yes, going slow and getting to know one's human and natural histories is essential to living well on a place.

The Natural Connection Between Food and Open Space

Do ranchers qualify as a keystone species based on their ecological, economic, and cultural importance? Ranchers, working viable ranches that sustain ecosystem services and contribute to the social fabric and local economies, are critical to a West that works. In addition to providing other vital services, they comprise an essential component to an intact rather than a subdivided West. Whether the land that is now in ranching remains there or shifts to other uses, we are up against the same need: to keep it open and unfragmented.

America is gradually waking up to one consequence of our globalizing economy: the loss of locally produced food from private lands that provide critical ecosystem services and open space. As ranching diminishes in the West and agricultural jobs move offshore, so too does the opportunity for our urban publics to reconnect with the rural tasks of husbanding food on well-stewarded lands.

Interestingly, these fragile relationships even relate to homeland security. When viewed in the light of rural and urban America, our government's concern over "Homeland Security" misses the most important point. A homeland's security is not simply based on military might. Home, land, and security blend together when urban people realize that ecologically sustainable food production is possible and that rural cultures matter—and when urban people are prepared to compensate farmers and ranchers for a healthy food product as well as for protecting open space, wildlife habitat, and ecosystem services. Gary Nabhan captured this when he wrote: *"The simplest fact about Western ranches is the one most folks tend to forget: raising range-fed livestock is one of the few economic activities that produces food—and potentially ecosystem health and financial wealth—by keeping landscapes relatively wild, diverse, and resilient."*

Imagine a time when urban people no longer take for granted the societal services of local farmers and ranchers, and Westerners eat locally produced food raised on private and public open spaces, offered and received with grace and at a fair market value. Equally important to this winning equation are rural people who acknowledge the importance of urban areas and offer a friendly handshake to their city neighbors. Perhaps, in creating this reality, we need to remind ourselves that humans, whether rural or urban, can be keystone species or the ultimate weedy species, depending on their relationships to land, and to each other.

Richard L. Knight is a professor of wildlife conservation in the Department of Forest, Rangeland, and Watershed Stewardship at Colorado State University in Fort Collins, Colorado. His interests focus on the nexus of land use and land health in the West. He can be reached at: knight@cnr.colostate.edu

Squaw Peak at sunrise

Fountain Creek

Alfalfa

Purgatoire Canyon sunrise

Sunrise ignites the sky, painting it with streaks — then swarms — of incandescent peach, pearl, orange, lavender, and, finally, a blue so strong it hurts your eyes. Then morning goes to work on the waves of sandstone bluffs, lighting them up to a deep brick-red, an intense red, but not quite as volatile a red as the branding irons being heated down in John Doherty's corral.

The Bar, the Triangle, and the L are each being turned in the propane flames by Uncle Bob, a retired eighty-four-year-old veterinarian from Fort Collins who barely knows Doherty, but is happy as hell to be here helping out. That's pretty much the way everybody feels. Some people Doherty only recognizes by their faces; names elude him. Some have come from fifty, sixty miles away, towing their horses in trailers, ready to extend the courtesy to Doherty that they know he'll extend to them when it comes time for their branding. Tomorrow it might be time to help Steve Wooten — Doherty's nephew three miles up the road — or maybe Kelly Bader or Mack Louden or whoever. But today is Doherty's day; some 228 calves will be branded with the Bar-Triangle-L. If they're males, they'll be castrated as well, most of them by Doherty's sure, quick hand. And somewhere between the start and finish, everybody will be treated to a solid lunch featuring . . . what else?

"Let me tell you, you *always* serve beef at these lunches," says Joy Wooten, Steve's wife and still — even after thirty-two years of marriage — his sweetheart.

"You want to see your crew evaporate, just serve chicken," says Steve, finishing Joy's thought the way she frequently finishes his. Then, with a laugh, he adds, "After a branding, there's no tofu, no cabbage burgers with a light sesame vinaigrette. Uh-uh."

"This is just the best time of the year," enthuses Joy, eyes flashing like blue fireworks, as usual looking way younger than her age, which, by the way, is fifty. "Baby calves, fellowshipping with neighbors, and springtime."

All around, the land spins out, full of texture and color and beauty. They're in one of the canyons carved by the Purgatoire River. At least out here it's called the Purgatoire. Head north, toward La Junta and such, and chances are you'll hear the locals call it the Picket Wire because that's what it sounded like coming out of the mouths of French trappers a couple of centuries ago. But whatever you care to call the river and the network of rugged red canyons it carved, it's hard not to reach the same conclusion as Steve Wooten.

"This is God's handiwork," he marvels. "It's just incredible out here. It's not very often you get to live and work in such a beautiful place."

Wooten stops. Looks around.

"You know, God still owns it, but we love it. We care for it." And after you care for it, put your hands in it, drag your foot across it; after "it gets to your soul" and "you get entwined in it" and "you *feel* it," and "when it's burned up and dry you feel terrible for it," and "when it's healthy it's beautiful," and you see how it "never looks the same each day," well, then you can't help but know "the land is alive."

And eventually, if you're John Doherty, and you've lived so intimately with and on God's handiwork for all of your sixty-five years of life, and you've bet your life and the lives of your family on its ability and willingness to grow grass to feed your cattle, you don't have to think about your answer when you're asked what the land means to you, because you've known the answer all along. You just have to find the right words so folks will understand.

So you look out around you and then you lean your head back and what you come up with is the only thing you can: "Oh, it means . . . *everything*."

Just like it did for Papa Joe.

Eighteen-year-old Joseph Doherty got off the train in 1879 and stepped right into Elmora, New Mexico, ready to conquer the world.

And a strange world it was, too. Not lush and green like his native Ireland, but high desert. Vast. Open. A place where an ambitious, hard-working lad could make something of himself. And so he did.

Joe became a deputy sheriff. Then he went into the mercantile business, opening a store. Then he got into trading cattle and sheep. But mostly he got into land. A lot of land. Shrewd and visionary, Joe weathered the Great Depression far better than most, adding to his property, which ultimately stretched out of New Mexico and into southern Colorado. Fifty thousand, sixty thousand acres of land — more acreage, he once boasted, than there was in all of Ireland.

By then, he had become Papa Joe, patriarch of a family. The land was passed down accordingly, eventually reaching the hands of John and Betty Doherty, Papa Joe's grandchildren. Both had been suckled on the outdoors — their mother, Lorean "Tiny" Doherty rode a horse until she was past seventy-five — but John maybe a little more so.

Sure, when the family moved north, near Kim, John became a smooth basketball player, even leading the tiny K-12 school to state for the first time ever. But he didn't want to play basketball in college. He wanted to compete in rodeo, to ride, and rope calves. He went to Sul Ross State in Alpine, Texas, sometimes referred to as "Possibly the most underrated little university west of the Mississippi." When he came home, he was ready to ranch because ranching was what he knew and loved best.

Betty, four years older, strayed from the land for a while. Married a structural engineer and moved around. Florida. Washington. Colorado. New Mexico. Her oldest child, Steve, was born in Las Cruces, but by the time he was in high school, the family was living in the suburbs of Denver. For Steve, however, the fit wasn't right. He didn't like the indoors so much. Too cramped. Could hardly focus for eight hours at a desk. Every part-time job that he had liked had been outside; sky was the ceiling he preferred. But it wasn't just being outdoors that felt so good, it was being outdoors on the family ranch near Kim. All the summers his family had spent down there, on the land, well, he couldn't let go of the memories. Didn't want to let go.

During his nephew's senior year of high school, John Doherty got a call. Could Steve come down and finish school in Kim and work for him? John was "just tickled" that Steve seemed so full of love for the land and the ranching life. But he wasn't an impulsive man; he did things in a measured way. And he wanted his nephew to do likewise. So he made him promise to go to college for at least two years. Then, and only then, could Steve have a full-time job.

Steve came down and joined a senior class of eight. Pretty soon, he had his eye on this really cute girl, Joy Jackson, who lived in Villagreen. She noticed him, too. How could she not? Being from the big city, he sort of stuck out. Take music. She liked Loretta Lynn; he introduced her to Pink Floyd. Her father thought she'd been "corrupted." Actually, she'd fallen in love.

Two weeks after Joy graduated high school, she married Steve. She was eighteen. He was nineteen. But age didn't matter. They both knew they'd found the one.

Steve took his uncle's education dictum one better. After community college, he went to Colorado State University to study animal science. Before he picked up his degree, he picked up his first daughter when Niki was born. Two years after that, when Uncle John made good on his promise and Steve and Joy were working on Red Rocks Ranch, Arin was born.

The girls were natural-born ranchers. How could they not be, what with their parents zipping them into their coats when they were scarcely months old and taking them for horseback rides? The Wooten daughters were competitive, too, like their daddy, and sports was a natural outlet. If it wasn't high school basketball, it was rodeo competitions. The modest ranch house became festooned with photos of the girls riding and roping, so many photos you couldn't be sure if some rooms really had walls. Both girls married young. Not as young as their mother, thank goodness, because she "would've died if either of my girls did that," but young enough. Didn't take all that long for both to have babies. Niki had Avery in September 2007. Two months later, Arin gave birth to Bray, which she thought sounded like "a really good cowboy name."

By then, both girls had moved away. Niki to New Mexico, Arin to Nebraska. Steve and Joy had lost their "cheap child labor" and the ranch had lost some of its youthful sparkle. Still, the parents were glad that their babies had married ranchers. They were glad that at least twice a year, they were guaranteed visits. One was Christmas. The other? Why branding time, of course.

The calf has been roped around its hind legs and plucked out of the small, bellowing herd. It has been flanked — pinned to the ground, one rancher pressing down on its shoulders, the other immobilizing its thick rear legs. In this case, the two ranchers are Arin and Niki. Meanwhile, another rancher — Art Dowell, the Kim School superintendent — is preparing to vaccinate the animal against Bovine Rhinotracheitis, parainfluenza, and a passel of other possible diseases. The branding irons are glowing; Kelly Bader walks over to the propane oven and accepts one from Uncle Bob. John Doherty has his small but very sharp knife — as sharp as Buddy Feemster can make them, which is sharper than anyone else can — ready.

If you use your imagination, you can freeze the scene for just a second — but only a second. Then, that stalled image is jump-started to life and begins rolling nonstop. The injection first, into the left shoulder. The three branding irons next, over the left ribs — Triangle, L, Bar — searing hair and flesh, sending a flurry of gray smoke into the sky. Then, because this one's a male, the castration. Then the Wooten girls are "spinning" the calf, letting it up and directing it back to the herd. Maybe forty seconds have elapsed.

You'd think there would be a frenzy, or maybe just an urgency, to the branding procedure. Instead, it's strangely calm, ruled by a quick syncopation, virtually no talking. Not so much an assembly line as a cowboy ballet—rope-flank-vaccinate-brand-castrate-spin, rope-flank-vaccinate-brand-spin—where everybody seems to know the rhythm and the steps through countless rehearsals. Maybe even because of cowboy DNA.

Or maybe it's something else. A sense of being loyal and proud and keeping a flame of tradition alive, same as the propane oven keeps those irons red-hot. Cowboys used to say you "ride for the brand." In other words, you take a rancher's money, you give him your loyalty, your commitment to doing the job the right way. Nobody's getting paid today, but that doesn't matter. The loyalty they feel to John, to each other, can't be bought anyway.

Thirty-eight calves are branded in less than an hour. Time to head up to the pasture near Courthouse Rock, where 120 more are waiting, along with maybe twenty other ranchers, ready to help. Camped on the hoods of trucks are sweet rolls, sticky-delicious, along with coffee strong enough to strip the enamel from your teeth and good enough that you don't care. A rancher named Gail Allen walks by, points to his very-full mug, and says with a laugh, "Know why I filled it that full? 'Cause it couldn't hold any more!"

The ballet starts up anew. Only now, four calves can be handled at the same time because of the work force. Different faces, same precision. The one constant is Doherty. He's the one who moves the people around to different jobs, from roper to flanker, flanker to brander. He does this more by gesture than words, pretty much the way he is all the time. He's a man respected for what he says, not how much, a man not given to soliloquies. A man appreciated for the tomes of ranching knowledge meticulously filed away in his brain.

"He has the eye of the master when it comes to making selections and decisions about his cowherd," says his nephew, Steve. "He has developed a cowherd that is better than a lot of pure-bred breeders have." It was the same way Doherty was when he was raising quarter horses; the same way he is today when it comes to studying range management, learning how to survive in what Wooten calls "this semi-arid ecosystem" where mercy isn't really part of nature's vocabulary.

Purgatoire River

Chacuaco Creek

Up here on Mesa de Maya, where you can see forever—and forever looks beautiful, by the way—Bob Patterson is folding his arms across his chest and not saying much because he knows he's standing on sacred ground. Right now the breeze is soothing and the sun is shining so bright it feels like it's squatting on your shoulder, but don't be fooled. The Mesa is about beauty and bounty, but it's also about no mercy. It gives, but it also takes. The wind can blow like a sonofabitch, the clouds can go from cotton to charcoal in nothing flat, and lightning can explode so big and loud it'll bring religion to an atheist. And, sure, the right grass will nourish your cattle, but the wrong grass will turn them pure crazy before it kills them. Maybe there's a little bit of the badass lodged in the Mesa's soul. Maybe that's why legend insists it used to be a place where an awful lot of outlaws chose to congregate.

But whether you're talking blessings, curses, or folklore, it doesn't matter. All of it is magic to the seventy-one-year-old rancher.

"It's kind of hard to put this into words because it's more of a feeling, y'know?" says Patterson, still looking out at forever. "But, I've drove this road here hundreds of times and it seems like I always see something different; something's changed. I been down here forty years and I still come across places and I'm figuring, 'Damn, I'm sure I been here, but this don't look familiar.' The land, it's just a feeling inside you; you just learn to grow to love it. That's why it needs to be preserved forever. Because it's sacred."

He turns and looks at his companion, almost as if he's looking for an answer. He doesn't have to look long. "Yeah, I think maybe you get it."

He starts walking toward his pickup. He'd like to stay and look out at forever longer, but there's work to be done down below on another part of his ranch, the one that spills off the Mesa and runs up against his son's spread and then his daughter's. Besides, Bunny's probably home from school now, and even after fifty years of marriage, she possesses another kind of magic that still casts a spell on him.

You could say that Bob Patterson became a rancher because of a tornado. Sure, he'd first toyed with the notion at age four, when he went to his first branding. And that pair of shiny black cowboy boots an uncle gave him a little while later didn't hurt any. But it wasn't until three years later, when his mother married Roy Eikleberry, a Baca County farmer and cattleman, that the deal was cinched.

Roy had arrived in Colorado in 1913, a nine-year-old boy whose family had lost pretty much everything in a Kansas twister. Ed Eikleberry, Roy's father, figured that was some kind of a sign. So he loaded up the family in a covered wagon, told Roy to mount up and drive their small herd of cattle behind them, and headed to southeastern Colorado, where he began homesteading. When Roy reached manhood, he went out and purchased his own place on the border between Baca and Las Animas counties. Nearest town was Utleyville, although the only reason it was a town was because it had a post office. Didn't have much else. This was where Bob Patterson came of age, growing up amid the cattle and the horses and the short-grass prairie, which was just about all he needed.

Along about this time, as Bob was learning how to ranch, up north in the Red Feather Lakes region near Fort Collins, a first-grade girl was reading a book. It was a story about a boy who visited his grandparents' ranch and got to do all kinds of interesting and wonderful things. Right there, the girl knew where she wanted to live when she grew up.

The girl's name was Burnette McCarthy, although everybody called her "Bunny," being as how she had been born right around Easter. Bunny came from a family of school-teachers on her mother's side, but her father started out as a coal miner. Years later, the man of her dreams would joke, "You know how Loretta Lynn was a coal miner's daughter and she became this famous singer? Well, with Bunny here, I thought, 'Man, I got it made.' Until I found out she couldn't carry a tune in a bushel basket."

The man of Bunny's dreams turned out to be Bob Patterson, settling in Fort Collins for what turned out to be a brief, one-year stab at higher education at what was then Colorado A&M University. As soon as they were introduced by respective family members, Bunny's heart did "a little flip." She went to college for a year, but when "Mr. Wonderful" decided he wasn't born to be a student and had better join the Navy before he got sucked into the Army, Bunny quit the University of Northern Colorado and married him so they could be together. Best move either one of them ever made.

Bob did some traveling in the Navy, passing through the Straits of Magellan, which, to his way of thinking, were "the toughest seas in the world." Even got to see the bottom of that world—the South Pole, a real "snow desert." But it was high desert that he longed for. When his hitch was up, he and Bunny came back to ranch in Colorado, two people sharing one dream.

In short order, two became five. First to arrive was a son, R.C.—for Robroy Carl, named for both his grandfathers. Then two daughters, Kisti and Kandi. Hey, Bob, what's with all the K's?

"Well, the idea for naming Kandi happened this way. Bunny was having trouble coming up with a name. She didn't like 'Candy,' though. Anyway, I was sitting on a tractor and Bunny was about to domino and—"

Whoa—domino?

"Yeah, when a cow's ready to drop a calf, we say she's about to domino."

So, Bunny, how do you feel about the arrival of your third child being described like that?

"Ranchers' wives are frequently compared to cows. We soon learn those are terms of endearment."

Right. Okay, where were we?

Oh yeah, Kandi. Well, apparently Candy with a K and an i was just fine with Bunny, so when she finished with her domino'ing the naming hurdle got cleared pretty easy. Meanwhile, as Bob and Bunny were expanding their own personal herd, Roy was doing the same with his, placing them on the 5,000 additional acres he had purchased in 1967. But cattle weren't the only creatures arriving on Roy's new property, located about eleven miles out of Kim. Bob and Bunny moved their family into an old stone ranch house there, and settled in for the long haul.

Kim was one of those now-you-see-it-now-you-don't towns. One store, one restaurant—actually they were one and the same—one church and, well, you get the idea. All three kids graduated from the Kim School, a K–12 institution that was built in the 1930s as part of a WPA project, the blonde sandstone hand-cut from a quarry and lugged into town on horse-drawn wagons. Bunny, even without her college degree, went to work as the Kim library teacher. At least she was the library teacher when she wasn't helping out on the ranch or, for one nine-year period, in the Kim Outpost. She and Bob bought the restaurant-cum-general store (Bunny jokingly called it the "Kim Mall") and worked like the dickens to keep it operating.

But helping Bob run the ranch and the store, and working at the school—did we mention she drove the school bus for a while to help make ends meet?—wasn't enough for Bunny. About a quarter-century after she dropped out of college, she decided to go back. In fact, she went to the same college at about the same time as her children—Oklahoma Panhandle State University, or as Bob called it, "Tumbleweed Tech." In 1984, after a number of intense summer-school sessions—after all, she still worked for the Kim School as librarian—Bunny got her bachelors in elementary education and library sciences. Over the years, "Miss Bunny"—as all the students called her—would add the title of technology teacher to that of librarian.

The Patterson kids went off, as kids usually do. In the case of R.C., it was to a twenty-year career in professional rodeo, riding bucking animals, getting a little busted up but not too bad considering he was in a business that "sort of encourages early retirement." Kisti became an accountant and got married and had kids. Kandi followed in her mother's footsteps and became a teacher who—naturally—married a rancher.

While the kids were off finding their ways, Bob and Bunny kept on polishing their own dream. Not that everything was always so pastel. There was that one winter when it wouldn't stop snowing. The power lines went down and they were without electricity or phone service for fifteen days. No matter. All that did was teach Bob that "Long as we got a little propane, we can manage." Then there was the cruel dryness, droughts so bad that "You think, 'Is it ever gonna rain?' And you know it is, but you just don't know when."

Beside the forces of nature, there was the force of blunt-as-a-hammer ranch economics. Gas prices going through the roof. Vet bills and medicine costs for his herd following close behind. Price of calves dipping. So many expenses, such paper-thin profit margins, that Bob would sometimes scratch that white brush-cut that covered his head and say to Bunny, "I believe I had more money in my pocket when calves were fifty cents a pound than when they were a dollar-fifty a pound."

Not that he was bitching. Bob was never much of a complainer. In fact, one friend went so far as to call him "almost a Pollyanna. Things don't bother Bob. If something happens, it happens. You just fix it and you go on." Or, as Bob was likely to say, "I'm always thinking it's gonna be a good year."

And it always was. Even if the price of calves went ambling south while the cost of everything was stampeding north. Even if some of his calves up and died on him. ("If a rancher says he never loses a calf, he don't have but one cow.") Even if the sky was empty of moisture. Because whatever bad news came along, it was always trumped by the good news that Bob and Bunny were doing what they loved. As Bob would say, "Hell, I been retired for forty years. It ain't work if you love it."

And if it ain't work, why would you need to take time off? Maybe that's why in fifty years of marriage, he and Bunny took one vacation. Happened when the kids got together and bought them airfare out to Las Vegas for a four-day visit to see the National Finals Rodeo. It was fun, sure, but by the third day both of them could hear the ranch calling.

Besides, no vacation spot could ever offer as pretty a picture as the land Bob laid eyes on in 1988. Seems some rancher's elderly father had gotten lost up on the Mesa, so a bunch of neighbors saddled up and went looking for him. Bob had been up there before, but not a lot. After he got his eyeful and returned home, he and Bunny were sitting around when he told her, "You know, if we ever get a chance to, I'm gonna buy that place."

And four years later, damned if that isn't what happened. Fifteen thousand acres where you could see forever.

The dirt road up to Mesa de Maya is still covered with a thin skin of mud and slippery enough so that even Patterson's three-quarter-ton four-wheel-drive pickup is doing a little bit of slaloming. Of course, he's glad that the truck is moving at all. Back at his ranch, it took more than a couple of stabs on the gas pedal to get it to turn over. "Sometimes this thing don't like to start and I don't know why," Bob had said. Then, with a straight face, "It's only got 200,000 miles on it."

The road is wet because not two hours before, one of those end-of-the-world rainstorms fell out of the sky, the kind where it feels like "someone is just pouring it out of a pitcher." The kind where the lightning can turn deadly real quick. One of these storms once cost Patterson ten heifers; every year he can always figure on losing two cows to the heavens. And not just cows. One year, Kandi lost two prized horses that were grazing up on the Mesa. Lightning hit one and then jumped right to the other. Bob found both of them lying dead, the grass they were chewing still in their mouths.

The pickup keeps bouncing, making its occupants feel like a pair of pants inside a washing machine, but it's stubbornly climbing, determined to reach the 6,800-foot plateau. Patterson pays the familiar jouncing no mind as he explains that he and Bunny are pretty sure it's the largest free-standing mesa in the world—not attached to any mountains or foothills, just out there on the high desert all by its lonesome; fifty miles long and up to six miles wide, stretching all the way to Oklahoma and most of the way to New Mexico.

West end of Mesa de Maya

North Fork Poudre River

Ponderosa pines

The view south

Looking north to Pikes Peak

St. Charles River canyon

East Spanish Peak

Pasture below the Culebra Range

Great dike below West Spanish Peak

A fistful of platinum hangs in the black mountain sky, igniting a shimmering aura over the hay meadow and curling creek below. The land is quiet, mysterious, and peaceful; all silhouettes and shadows. It probably doesn't look much different from the way it did nearly sixty years ago, when a genteel electrical engineer from a Philadelphia suburb who had absolutely no business coming west decided to do just that. Piled his wife and three boys into an old Plymouth woodie and hit the road, finally ending up in a lovely valley about a mile from Clark, which was about twenty miles from Steamboat Springs, which was a million miles from anything in the engineer's comfort zone. It was here he set about learning how to be a rancher in bits and pieces.

Truth be told, there were some bits and pieces he never got the hang of because, in a way, he always stayed an engineer, a man more suited to sweeps of vision and big projects than cattle and haying. A man who dramatically changed not just the look but the very ethos of his adopted valley. A man of whom it was said, "If you gave him dirt, diesel, water, and electricity, he'd be in heaven."

And the person saying that ought to know. He ought to know because he is the youngest son of the engineer who came west. He's also the only one who fell in love with the land and the cattle and stayed to learn their ways; remaining on the ranch year after year, until the years became decades, and the decades tumbled past a half-century, and there is still no end in sight to the son's love of the land. You can see that love in his eyes, but even when you can't see, like right now, when his face is washed in moon shadow, you can hear it in his voice. It is a soft voice, a voice that cites Mother Teresa and the intricacies of cattle genetics with the same kind of curiously easygoing gravitas.

But, as the son rests on his deck and looks out at that backyard of creek and meadowland, there is really no need for voice at all. So he sits back quietly, ripples of calm moving out from him, seeming to create a protective buffer that guards him from the endless attacks by crazy mosquitoes. He is not a man given to anger, at least not frequently. Oh, he has a temper—you help raise three daughters, wrestle with nature for forty years, and expose yourself to the switchblade attacks of political campaigns and, damn right, your pilot light is going to flare sometimes—just not much of one. He doesn't have a lot of use for drama and has spent most of his life working on ways to head off crises because he'd rather solve a problem than rant about it.

Of course, at the moment, he seems as problem-free as a man can. Inside the small web of quiet that he has woven, he is content to watch the night wrap itself around his land, pleased by the thought that it will always be this way; it will always be grass and water and, hopefully, cows. It will be the way it was when that old Plymouth woodie finally stopped, and a two-year-old boy from a Philadelphia suburb first scampered across the wonderland that is the Upper Elk River Valley.

It will always be this way because of the beliefs—and the actions—of the quiet man who sits on the deck. Because, deep in his heart, Jay Fetcher knows that some things should never change.

The modest pickup truck has left behind the town of Clark ("Elevation 7271, Population ?"—at least that's what the sign says) and is heading up into the undulating beauty of the land that sits under the stony gaze of Hahn's Peak. Along the way it passes through small developments with names like Deer Trail Estates and Badger Meadows.

"We have a saying up here—you name your subdivisions after what you've destroyed," says sixty-one-year-old Jay, smiling easily-but-wryly beneath a bushy moustache.

Then he returns to talking about cattle. Or, more precisely, how his uncle—the one who came west to ranch with Jay's father—was the first person in the area to crossbreed Angus and Herefords. Although crossbreeding eventually became a common practice, back in the 1950s, Stanton Fetcher may as well have been mixing eye of newt with dragon tongue because he couldn't give away his calves. "Yeah, the Fetchers were ahead of their time in this area," says Jay, adding, "And they paid for it."

He drives past waves of yellow Mule's Ear, which flood the land like the poppies in the *Wizard of Oz*. He points to Steamboat Lake, whose water has buried what used to be 900 acres of his family's ranch and, along with it, the land where he learned his cowboy skills. Seeing the reservoir naturally triggers stories about his father's role in getting the dam that created it built, including the one about how he managed to fit the final pieces of the deal together over a poker game in New Mexico. It also leads him to make brief mention of Stagecoach Reservoir and how his father helped build the dam that created it, too. This leads to a mention of the Yamcolo Reservoir, on the Yampa River, which John Fetcher *also* helped bring into being. Then talk about these significant expanses of water takes the conversation in the direction of Lake Michigan, where his father, a Chicago native, learned to sail. This, in turn, shakes out a memory of long-ago family vacations in the Caribbean on a fifty-foot sailboat with his father shouting orders and everyone scrambling to follow them.

The truck eases around a corner and passes a very rustic homesteader cabin, wheezy and weathered, lacking electricity or running water. The son explains this is where his father still comes every weekend he can, just to get away. John Root Fetcher is ninety-six now, but he still works—as a district manager for the Upper Yampa Water Conservancy District—still plays tennis—doubles, twice a week—still looms large in his son's life. Maybe he isn't quite the irresistible force he once was, the man who was "someone you couldn't say 'no' to; he just wore you down," but he's still a presence, apt to emerge in conversations that Jay has with friends and strangers as someone still very much inclined to giving directions and expecting those directions to be followed.

If the old man can be relentless about having his way, well, he is equally relentless about doing things the *right* way. Take the red, vintage Mobil gasoline pump that stands over a 1,000-gallon tank on the Fetcher property—the pump that's not just an antique, but a working antique. Four or five years ago, the glass globe that sits on top of it was broken. The father insisted that the replacement be another antique globe. Same vintage, not some phony replica. Searched all over until he found one. Cost two hundred bucks, but it was worth it. "He just wants everything to be right," says the son, smiling at his father's perfectionism, but maybe wincing a little, too, maybe remembering what it was like to grow up with a man whose idea of a margin for error was zero, whose standards never deviated from painstaking precision.

But if there is respect in Jay Fetcher's voice when he talks about the man he was named after—yes, he's a junior, although hardly anybody knows that—there is something else in that voice, something that comes from a place deep inside him, when he speaks of his mother.

"In all this history that I am, I truly admire my mom the most," he says of the former Clarissa Wells. The way she came west with three sons to a strange, isolated place and was able to "adapt and survive and make her mark is truly remarkable, just remarkable." When he speaks about his mother, a Quaker of quiet yet unshakeable beliefs, his voice grows softer yet somehow more powerful. And when he says, "I'm more like my mom than anyone," it's impossible to miss the love and humility.

It's also impossible to mistake a son's pride in his mother's creative side when he shows you something he has copied out of her journal, a passage Clarissa Fetcher wrote on December 1, 1949, not long after arriving out west: *The high mountain peaks around us have snow and they gleam like diamonds in the morning sun and then take on lovely rosy reds in the evening sunsets. And the mornings are lovely. Honestly this valley, when the sun is just starting to come up over the hills in the morning is beautiful. The sun first hits the snow peaks and they stand out gleaming white in the dark dawn for a while; and then it breaks through a few dips in the hills and comes streaming across the fields.*

Maybe it was from his mother that he inherited what he calls his "sense of optimism." An optimism that discourages him from padlocking the gate to the 650 acres of Fetcher property known as Hahn's Peak Ranch, unconcerned about trespassers, vandals, or thieves invading and wreaking mischief, trusting in the better, moral side of human nature. Maybe it was from her that he took away the sense of good-natured charity that has led him to grant locals free Elk River fishing privileges on his land instead of leasing it off to some private group and pocketing another ten grand. Maybe it was from her that he gained the sense of whimsy that once led him to take a slightly different approach to filling out a personal information questionnaire from a newspaper when he was running for the Colorado Legislature. Asked to list his pets, he wrote "300 cows, 12 chickens, 5 horses and 1 cat."

Then the road curves again and the sweep of land changes and, as he gets out to show you around, Jay Fetcher doesn't seem optimistic or whimsical anymore. He is looking at death. Purple, gray, and rust-colored death. At least those are the ghost colors you see in the distance when you scan the thousands and thousands of trees that are dead or dying, victims of the pernicious Pine Bark Beetle. The line of death recedes and expands in so many directions it's hard to take it all in, it's hard not to be suffocated by it.

In the silence you hear Fetcher say—softer than usual—"It changes the whole environment." There is more silence. Then, "I'm trying to be accepting of it." More silence. Finally, "I think the hardest part is I'll never see it back the way it was. These trees are 80 to 100 years old. I'll be gone before they ever come back. I'll never see this the way it was."

He gets back in the pickup and continues driving. His mood brightens as he talks about hay and cattle and how ranching up at 8,000 feet has its problems, but he'll take them any day of the week compared to what other ranchers face. How, even when there's drought in other parts of the state, there's always water up here for cattle and irrigation. How, even when you have to bulldoze ground, it always comes back, grass, time, and the land's unremitting beauty melting the scars into submission. How predators are rarely a problem for his herd, largely because there's a generous supply of fresh mutton provided —unintentionally and unwillingly—by Jay's sheepherder neighbors. How "This land is really so forgiving."

No, he can't do anything about the beetles. Neither his will nor his hard work nor his meticulous planning can stem that tide of destruction. But if he can't do anything to stop this onslaught, that doesn't mean he feels powerless against all forces. So when another kind of threat was looming, when a foreboding change started to rise up in the valley, and it seemed like his heritage, his culture even, was under attack, Jay Fetcher didn't stand idly by. Instead, the son of a Quaker and an irresistible force rolled up his sleeves, spit on his hands and got busy.

Funny thing—or maybe not so funny—but Jay Fetcher might never have gone and helped pioneer the idea of conservation easements in Colorado if his father hadn't gone and created "a monster." At least that's what John Root Fetcher would come to call the ski resort he built in Steamboat Springs in 1963.

By then, Stanton Fetcher had left Colorado, and John, Sr., was looking for a new challenge, having engineered the hell out of the ranch—which, by the way, wasn't doing all that well. It had been a rough go pretty much from the start; the two brothers had been flimflammed by the realtor who sold it to them in 1949. He told them there was enough land to raise 500 cows and grow 1,000 tons of hay. As it turned out, there was enough land to raise and grow half of that.

Once he and several partners decided sometime in the late 1950s that Steamboat Springs needed a ski area, John Fetcher wouldn't let go. He mortgaged his ranch for the sake of the Storm King Mountain Ski Company. Sometimes, he'd transport cows down to the Denver Stockyards in the middle of winter and sell them so he'd have enough to pay his ski employees. He'd use ranch tractors to clear trails and pull cable; he drove his farm truck to San Diego to pick up parts of the chairlift. It wasn't unusual to see him—the president of the operation—skiing down the hill with a plunger in his hand to fix a toilet. After more than three years of engineering and round-the-clock activity, the Storm Mountain Ski Corp. opened for business in 1963. It wasn't exactly a gangbusters debut—opening day temperatures sat near minus-thirty, and one of Fetcher's partners accidentally banged into the car of the only paying customer.

acreage to what it had been before the reservoir had been built. And you're damn right he had his own opinions about things. Take herd size. The father thought they could be grazing fifty more cows than they were. The son didn't want to tax the land; didn't want to "stress the resources." He decided "sustainability" didn't just apply to the land nourishing his cows and growing his grass, it applied to all the land; to the streams and the hard, barren ground. It meant caring about—and for—all of it. He came to think of himself as a "naturalist," a "pragmatic environmentalist."

Now all this might have sounded a little screwy to John, Sr., but even if he was a force who was used to getting his way, he was also someone smart enough to know a damn good rancher when he saw one. So the son prevailed on this issue. When it came to running the ranch, he usually did.

The years—calving season, fixing fences, haying season, calving season, fixing fences, haying season—fell away. The son continued to sail along, running the ranch adroitly, often guided by a cockeyed philosophy that went like this: "People say, 'If it ain't broke, don't fix it.' I say, 'If it ain't broke, break it. Then fix it.' "

One day, he stepped back and figured maybe he was the one in need of fixing. Here he was, working all the time, obsessed with a million tasks, isolating himself from everything but the ranch—*Damn, I'm turning into my father*—how was he ever going to meet a woman? How was he ever going to have a family of his own? He thought about chucking the ranch for a while and becoming a truck driver, just to, y'know, see the country. Meet interesting, new people. Meet *any* people. Then something wonderful happened.

Gael Mahony was in Steamboat to visit a friend before she took off to see the world. The friend happened to be married to Jay's college roommate. The roommate happened to be working on the Fetcher Ranch. Kismet. Gael and Jay met. Two weeks later they decided to get married. Five months later they did.

Molly arrived first. Then Kalley. Finally Annie. The girls all grew up on the ranch. They all learned to ride. They all learned to love the land. They were never required to work it. Oh, Gael and Jay had their disagreements over the fact that he never levied demands on his daughters. John, Sr., also wasn't afraid to weigh in. *His* kids had learned what a day's work was, by God. The ranch had put some steel in their spines, had forged them into productive adults, assets to society. How could Jay not mandate steady chores and responsibilities? Why weren't those girls out raking hay? Doing something? What was Jay thinking?

Gael sort of agreed. She thought the girls were lucky to have grown up on the ranch. She knew that ranching was a "dying industry" and worried about what would happen to the land after she and her husband weren't around. Hell, even hard-working Jay didn't want to be "fixing fences at seventy," she knew that. When it came to the ranch, maybe more should be asked of their daughters.

But Jay held firm. He remembered all the things he had missed because he had to jump off the school bus and onto the haying machine. He remembered all those Sundays when a day off sure would have been nice. He remembered being pounded by his father's work ethic and then pounded some more by the ranch's endless demands. He thought of how it was a good thing he loved the life because if he didn't. . . .

Sure, he was concerned about what would happen to the land after he was unable to work it. Or dead. You don't relinquish so much of your sweat and toil—to say nothing of most of a finger—to something and not think about what lies ahead for it, who's going to champion it. But he wasn't about to be overcome by fear of the future anymore than he was about to try and navigate his children's directions.

So if anybody asked him to explain what his thinking was, he could, because he'd done a lot of thinking about it. And what he thought was this: "This is a hard life. It has to be somebody's choice; it has to be something you want to do." And if you don't, well, "I'll be blunt: I love the ranching lifestyle. But I don't wish it on anyone."

The Big Picture? The future? That was his daughters' decision. He'd stick with helping them work on the small things. And he'd do that with great love.

Wearing a pair of torn clogs and really dirty jeans, Jay Fetcher lounges in a chair, looking out on that unbelievable backyard of his, talking in that pleasant, strong voice. The one that has no twang or rolling drawl or mention of a crick. A professorial voice, but one that is accessible, easy to listen to, easier to trust. Right now, the voice is talking about why the hell a soft-spoken rancher who is a Democrat living in an area where Republicans spawn like the mosquitoes that swarm around his deck would ever get involved in elective politics.

Which, by the way, is just what Jay Fetcher did. And not only that, he did it twice.

The first time it happened was 2000. The ranch had become "routine; I'd done it for over twenty-five years." He was looking for a new challenge. He had enjoyed his time on the local school board and local fire district board. He looked at the legislature. Colorado House District 56, to be precise. Hmm. He could pay someone to come feed the cattle. And now that his brother Bill was done with his Navy career, he could help with the haying and the upkeep of the machines. Geez, Bill was a perfectionist—just like their dad—so meticulous, he'd wax the tractor to keep the hay dust off it. Yeah, if Bill could help then Jay would figure out a way to handle the rest. After all, hadn't he got the ranch on such a humming schedule that he spent "barely five percent of my time on crisis management"? Running for office. Hmm . . . .

It was the biggest risk he had ever undertaken. And when he lost by 160 votes, he was "devastated." But apparently not cured. Of politics, that is.

In 2004, the Democratic Party came calling. Republicans held a one-vote majority in the Colorado Senate. Let's see, in which districts might they be vulnerable? Who might have a good shot at ousting a GOP incumbent? Well, Jay Fetcher might. He was well-known. He was moderate. He was popular. He was one-hundred-percent rural—hadn't he gone to school in Clark's two-room schoolhouse? Hey, Jay, asked the party brain trust, feel like taking on Jack Taylor in District 8? Sure, he's the incumbent. Sure, he's angling for his seventh term in the legislature. But you can beat him, Jay, we know you can.

For Fetcher, it wasn't a hard decision. From his standpoint, there still weren't enough ranchers in the state house, still weren't enough voices that could speak out on ag issues and have their words come from the heart. And if the odds were long, well so what? It was like Mother Teresa once said, "God didn't ask us to win. He just asked us to try."

Things got ugly fast. Fetcher was accused of voting against a bill to expand I-70 when he was in the state legislature. Only problem was, he had never been in the state legislature. Fetcher was accused of "cutting a deal" that would up putting control of Western Slope water into the district of a Boulder congressman—which is only about the worst thing you can say about someone from the Western Slope. That wasn't true, either. A flyer was circulated that had a picture of Fetcher and National Democratic Party Chairman Howard Dean. Dean looked particularly crazy in the photo and the language accused Fetcher of being financed out of "New York and California," and implied he favored higher taxes, opposed private property rights, and was all for gun control. They might have added that he stomped on the Bible and beat his dog, but there wasn't room.

And yet, when the mud congealed and the dust settled, Jay Fetcher did all right. He even won Routt County—where both he and Taylor lived. In fact, "we kicked his ass by 20 points," he would say, not sounding very Quaker-like but pretty damned pleased just the same. But what happened in Routt wasn't mirrored in the district's other five counties. When the counting was over, Fetcher had fallen short of Taylor by 1.5 percent. Some people would call that a moral victory. Some people would call that a loss. Some people would figure that two elections was enough brain damage. Jay Fetcher was one of those.

"Hey, Gael," he calls out to his wife, who has joined him on the deck, "Am I ever going to be allowed to run for office again?"

"No," she says, and they both laugh.

They are an interesting couple—the city girl from Denver who is vivacious and has just a hint of urban swagger, and the rural kid who grew up more comfortable around cows than girls—powerfully bonded, yet clearly not joined at the hip. Easy to be around. Even after thirty-six years of marriage, three daughters and three grandchildren (and lord knows how many calves), they effortlessly demonstrate the chemistry that instantly drew them together and put a stop to Jay's plan to become a truck driver and Gael's wanderlust.

"She liked the view, I think that was what made her stay," laughs Jay.

As far as he's concerned, it was Gael who helped bring him out of his own skin, helped him to be less of an introvert. Odds are he's right. Spend a little time with Gael, listen to her breezily describe her first encounter with Jay ("I knew a good thing when I saw it"), experience her warm but won't-take-no-for-an-answer hospitality ("Oh, you have to stay for supper; it's just left-overs, but they're pretty good"), and you might begin to understand how she could siphon off some of a man's shyness. You might also get an idea of how her outspokenness could rub off on someone when you hear her talk about the people she calls "deep-pocket ranchers." Only they aren't *real* ranchers. "You can be a garbage man or an orthopedic surgeon but then come up here and say, 'Hey, I'm a rancher,'" says Gael, adding emphatically, "But you're not."

Sure, it used to drive Jay nuts when he'd see ads for "35-acre ranches." But maybe it was because of Gael's influence that he actually began to call up the real estate companies and say, "That's just wrong. Thirty-five acres isn't a *ranch*."

Because he knows that it isn't just acreage or livestock that turns land into a ranch or a man into a rancher. No, for that to happen, you need something else, something beyond possessions. Time was, people helped each other in the valley. You needed to borrow a man's tools or machines, he loaned them to you. You needed to cross his property with your cows? Why, sure. You worked together. You worried about the weather together. You were part of a community that was built on more than property values and fancy houses.

Now?

Well, take the guy who bought a ton of land near Jay and Gael. He doesn't live there, mind you. Just comes up every now and then. And when he does, says Gael, "He's not here to *be* a neighbor. He's here to get away from neighbors." He even tried to block the Fetchers from gaining access to his property. Jay and Gael had to hire lawyers to prove that, yes, indeed, they had a legal right to take their cattle across his land. *Lawyers, for chrissake.*

It's happening more and more that people come here and think they're living in a vacuum. Like the folks, says Jay, who built one of those "starter castles." Now that would be okay, except they built it high on a ridge where everybody can see it sticking up in the midst of all that natural beauty. "How selfish is that?" asks Jay, adding, "And, what's worse, there's never anybody up there. The only time you see any lights on is at Christmas."

Then he stops. It's as if he's reminded himself—again—that it's pointless to worry about these kinds of things. You just have to let go, just like you learn to do without a chunk of your finger. You learn what you can do and what you can't do. You learn that as much as you'd like to stop that terrible plague of beetles and save those magnificent trees, you won't be able to. You learn to walk away from that, just like you learn to walk away when all the votes have been counted and the other guy is elected.

You learn that, ultimately, God doesn't expect us to win, just to try.

So maybe Jay Fetcher's children will not become ranchers like he did. That's okay. Remember—you don't do easements for your kids, you do them for yourself. You do them so the land will remain unscarred by starter castles or golf courses or people who have the audacity to call thirty-five acres of ground a ranch. You do them so the land will stay rough, but very real, ready to produce. So it will stay just as it is now, with livestock and trees and whispering hay and a curling creek all right in front you, getting ready to turn into silhouettes and shadows. So you can sit and watch all this and tap your heart and feel a certain kind of light inside you flare up, glowing like a fistful of platinum in the black mountain sky.

[John Root Fetcher died Feb. 6, 2009.]

WEST SLOPE
ROCKY MOUNTAINS
FOOTHILLS
GREAT PLAINS
VERMILLION
FETCHER
Walden
ROBERTS
EAGLE ROCK
SWEDLUND
Sterling
Maybell
Craig
Steamboat Springs
Ft. Collins
MANTLE
PARKVIEW MOUNTAIN
R&R
Ft. Morgan
WRAY
Wray
Meeker
RUSSELL
Kremmling
YUST
PINEY PEAK
Vail
Dillon
Denver
Eagle
Glenwood Springs
Rifle
HILLSIDE
COLD MOUNTAIN
DEER VALLEY
LOST MARBLES
HARVEY
Aspen
Leadville
Limon
Burlington
DAKAN
AUBERT
Grand Junction
VOLK
Buena Vista
Colorado Springs
BRETT GRAY
ESCALANTE
Delta
SUNRISE CANYON
COGAN
BAR OPEN AK
FOUNTAIN CREEK
TRAMPE
Gunnison
Salida
Montrose
IRBY
Pueblo
CENTENNIAL
HILL
RIVERGATE
FLYING X
RUSK
3R
LAST DOLLAR
Ouray
Lake City
BEAVER MESA
Telluride
Lamar
La Junta
SCHMID
WILSON
RIO OXBOW
South Fork
Monte Vista
Walsenburg
NOTCH
WEMINUCHE
SAM CAPPS
CANTRELL
REDBURN
Alamosa
BEATTY CANYON
JE CANYON
Springfield
Cortez
Durango
Pagosa Springs
RIO PINOSO
HARTONG
CROSS ARROW
SALAZAR
Trinidad
PATTERSON

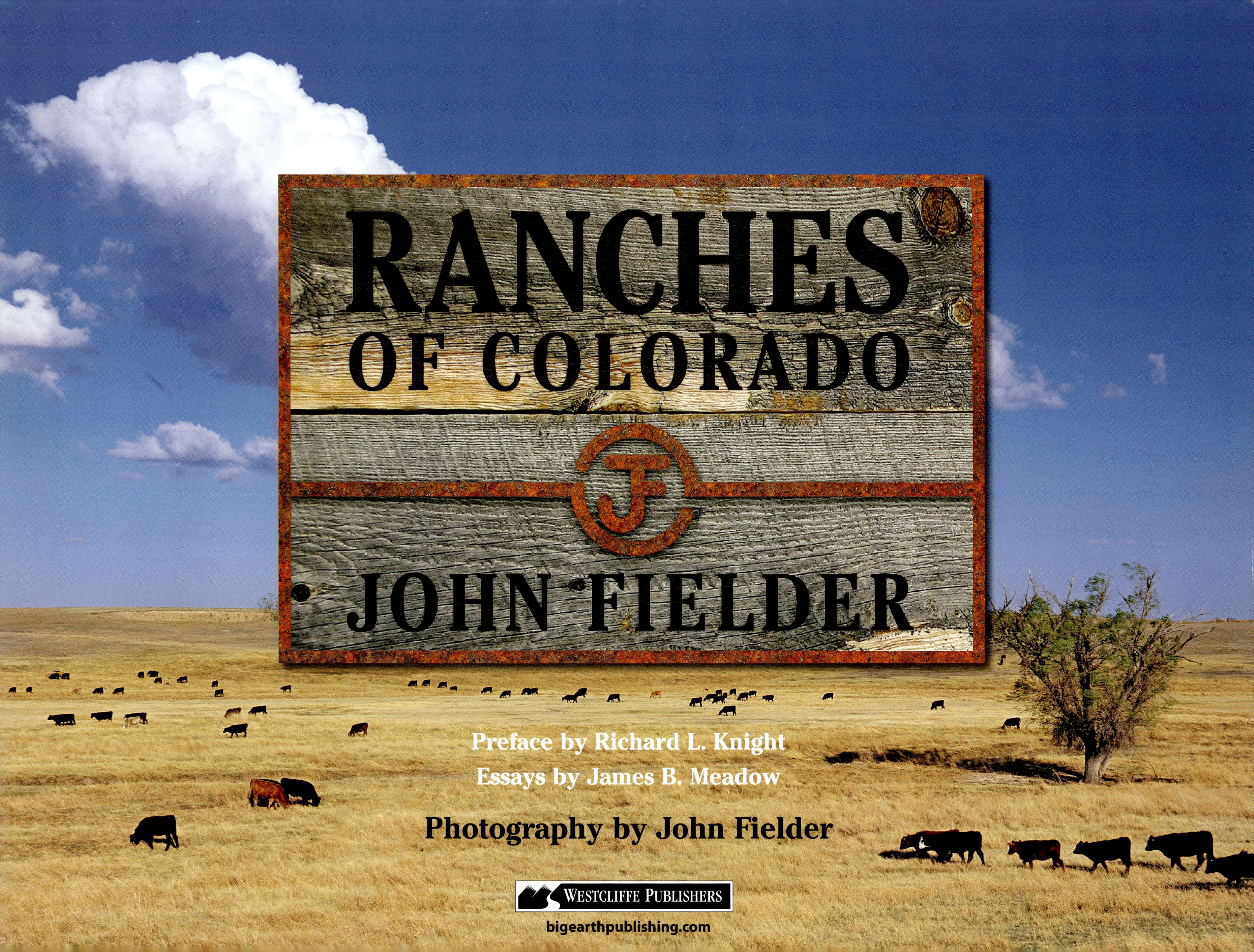
RANCHES
OF COLORADO
JF
JOHN FIELDER
Preface by Richard L. Knight
Essays by James B. Meadow
Photography by John Fielder
WESTCLIFFE PUBLISHERS
bigearthpublishing.com

The Conversion of an Environmentalist

John Fielder

During my forty-some years taking pictures of Colorado, I've placed my tripod against many a barbed wire fence and gazed into the Colorado ranch meadow beyond. This was usually just an unplanned stop along a highway on the way to a prescribed week of photography in a beautiful wilderness area—just a way to fill the files with an extra photo. I have spent most of the past thirty years making my living by documenting Colorado's public lands—national forests and parks, rivers and canyons of the BLM lands, state and local parks and open space. For good reason: Since the Wilderness Act of 1964, Americans have fought to protect our remaining wild places against mounting odds, and photography has been a valuable tool for me and others to show what's at stake. America has done a pretty good job protecting the best of our public lands from development, however, over the years I've noticed dramatic changes within the bounds of those barbed wire fences—and not for the better. Colorado's private lands are being subdivided, vivisected, at an alarming rate. So, in 2007, I decided to point my lens in their direction.

My first visit to Colorado was in the summer of 1964 on a school field trip from my home in North Carolina—we camped in Rocky Mountain National Park for a couple of days before heading up to British Columbia. My first full summer in Colorado was in 1967. My Uncle Fred had just moved from Chicago to take the position of CEO for steel manufacturer Colorado Fuel and Iron in the southern Colorado town of Pueblo. Knowing that my 1964 visit to Colorado had been something of an epiphany for me—at age fourteen I was smitten by the grandeur of 14,000-foot snowcapped peaks, so much more impressive than my tree-covered Appalachian summits—he set up a telephone interview for me with a rancher friend of his for a ranch-hand job in the Wet Mountain Valley, an hour's drive west of Pueblo. I was hired, and on June 1, 1967, I found myself standing beneath the Sangre de Cristo Range on a 400-acre Hereford-cow and quarter-horse spread.

The Sangre de Cristos are no ordinary Rocky Mountain range—they contain several "fourteeners," peaks above 14,000 feet, all but one of which I could either see from the ranch or from a short ways away. They are one of the longest mountain chains on the planet—they begin near the Colorado town of Salida and end 200 miles to the south near Santa Fe, New Mexico. The Wet Mountains lie east of the Sangres, are not nearly as high, and are best known for the massive amounts of silver extracted from them in the 19th century. In between the two ranges is the Wet Mountain Valley which, in my opinion, is the most beautiful ranching valley in the West. Hay meadows extend as far as the eye can see from north to south and east to west, and are interrupted only by the tiny towns of Westcliffe and Silver Cliff, the latter a remnant of mining days and the former a place for ranchers to gather. Remarkably, not much has changed here since 1967. Transplanted city-dwellers have built homes in the foothills, and tourists love to stop by, but the valley floor is still populated by more cows than people. This wonderful place was my home during the summer of my junior year in high school, and what a summer it was: City slicker kid from Charlotte, North Carolina meets a whole 'nother culture!

On the first day I got my own room in the ranch house—I also got a big black quarter horse just removed from the Pueblo quarter horse racetrack, along with a warning not to "run" him. This three-year-old stud was used to going in a straight line on a quarter mile race track, not doing what quarter horses are supposed to do on a ranch: "cut" cows out of a herd. My boss, Chet, was in the process of training him to be such a cutting horse; to stop and start on a dime and change directions without any help from his rider while chasing and directing a calf to a place the cowboy wants it to go. So the last thing Chet wanted me to do was run this horse in a straight line—which is exactly what I did when he left me alone on the ranch for a weekend. I have never forgotten the exhilaration of galloping that horse at top speed across the meadow—the first time in my life to do such a thing, and almost the last. My boss Chet got home on Sunday night, stopped by the barn to check on things, entered the house, looked me in the eye, and asked me if I had run the horse. I admitted that I had, was told never to do it again, and was docked my next Sunday day off as punishment. Later, I figured out that Chet had noticed the salt caked up on the horse's neck from his run across the meadow. Stupid me for not washing him off.

For almost three months, I learned the ranching life. Up at 5 a.m.; big breakfast; to work by 6; big hot lunch, which is called "dinner," at noon; back out an hour later; home for leftovers at 7 p.m., and in bed by 9. This was the routine six days a week. I got Sundays off, which I usually spent with Chet's son Mike, breaking into old mine buildings in the Wet Mountains to look for "girlie" magazines. We got caught once, but only after being chased by Sheriff Stan Depriest for an hour through a long arroyo, thinking the whole way that we could escape the law. How dumb of us to think that we could outsmart a grizzled western sheriff. Chet picked us up at the Westcliffe jail mad as hell, resulting in the loss of another Sunday off.

Part of learning the ranching life was learning how to drive: D-2 and D-5 Caterpillar tractors to move dirt around and to plow manure out of the corrals; farm tractors for various purposes, including just getting around the ranch on something other than a horse, a job done today by four-wheelers; a winch truck that we used to raise the trusses of a barn we built; and various pickup trucks and cars, including the one Mike and I were given for our days off, a 1956 Chrysler Imperial with no muffler that could be heard from one end of the Wet Mountain Valley to the other.

I also learned how to ride that fast horse pretty good, albeit at slower speeds while he was being trained as a cutting horse. When Chet was away for another weekend and I was left in charge, I got a call Saturday night from the ranch next door, informing me that our cows were out on the county road. At 1 a.m. I saddled up my horse in the dark, rode across the hay meadows, herded the cows back into the pasture, and fixed the fence under the light of a half-moon. I built an entire corrugated steel barn with one other man, and I helped manage water flowing through the ranch ditches used to irrigate the hay meadows. By summer's end I learned how to cut and bale hay, and even worse, to pick up and toss ninety-pound bales onto a trailer since we had no mechanical hay stacker. Nevertheless, this got me in shape for summer football practice, which would start as soon as I got home to North Carolina. Little did I know then that my experience as a Colorado ranch hand would serve me well when, forty years later, I would need to communicate with fifty "Chets" in order to complete a project about ranching in Colorado.

I worked two more summers in Colorado, prospecting around the West for gold, silver, copper, and molybdenum as a junior geologist for the non-ferrous metal exploration division of Uncle Fred's Colorado Fuel and Iron. Three days after my college graduation in 1972, I moved to Colorado for good. I managed an eight-year career in the department store business while developing nature photography and outdoor skills on my days off. In 1981 I left the corporate world and turned my avocation into the career that I practice today. Since then I have explored much of Colorado's public lands under the management of the Forest Service, Bureau of Land Management (BLM), and Park Service. In addition, I have photographed many Colorado state parks, as well as local parks and open space. I have explored Colorado by hiking, skiing, and rafting for thousands of miles, and traveled many more thousands by vehicle. I have witnessed moments of light and weather that defy imagination, and I have seen the fecundity of life as manifested by plants and wildflowers and migrating creatures large and small. I have experienced the medicine of sitting alone on a wilderness ridge at 12,000 feet with not a thought in my head, stimulated by just the sight of mountains all around me, the sound of wind in the blue spruce below, the aroma of alpine tundra, and the touch of a cool July wind against my face. I have published many books about these places and spoken publicly to thousands of people about the need to protect them for the sake of future generations and for the benefit of biodiversity—the multitude of life forms with which evolution has blessed Earth. I became an environmentalist because of the sublimeness of what I have seen and experienced. I learned that human health and economy depend upon the integrity of nature's billion-year-old processes, and that all living things are inextricably connected to one another, and to the places, the habitats, in which we live.

Along the way I recognized that Colorado is only a political subdivision, and that nature's subdivisions, which are called ecosystems, have boundaries too. In Colorado, elevation is the main determinant of these boundaries. Many plants and animals live their entire lives within a single ecosystem, yet some are not constrained. For example, elk and deer live high in the mountains in the summer, but retreat to river valleys for winter forage. One variety of larkspur wildflower grows in the foothills ecosystem, another in alpine meadows at 13,000 feet. I learned that for biodiversity size is everything, that large landscapes unfragmented by development, even by roads and trails, promote greater biodiversity than small ones.

I became disenchanted with people and constituencies who ignored the impact of growth and development on open and natural landscapes during land use planning processes at federal, state, and local levels. In 2000, I even led a growth management ballot initiative campaign in Colorado. Furthermore, I scrutinized industries that I felt did not have the best long-term interests of the land in mind, if not the general long-term health and welfare of Colorado's people and its economy. In doing so, I concluded that logging, hard rock mining, and oil and gas development, the so called "multi-uses" that had been allowed on public lands for a century, were bad for the natural environment—and not so good for the economy either, at least from a long term perspective. For example, hard rock mining made Colorado great in the 19th century, but the cost of the environmental cleanup today makes one wonder if what was gained was worth it. One cannot help but ask if this same thing will one day happen with today's explosive oil and gas industry. Surely the oil and gas will play out just like the gold and silver did. What will Colorado look like then, and what will we hang our economic hat on if we've destroyed the very reasons why people love this state? In the long run, I believe that Colorado can only protect those reasons by being an "attractive" not an "extractive" place. By predominantly selling our clean air, pure water, beautiful views, and recreation, I believe that we can only sustain a healthy economy and ecology. Tourists will always want to spend a pretty penny to gain a respite from urban life. If we screw up Colorado now for the sake of short term gain and short term jobs, it will be at the expense of the state's greatest resources.

At one time, I even scrutinized the very industry that had employed me in the summer of 1967: cattle ranching. While backpacking and llama packing into our public lands, I witnessed the destruction of wetlands that occurs when cattle and sheep are allowed unconstrained grazing in these most fertile of places. I read up on the history of ranching, and learned how much damage was done to the West's rangelands in the 19th century, when tens of thousands of cows were driven from one place to another, causing environmental damage from which, unbelievably, evidence still remains in places today.

Ranching: Good News for the Economy, Ecology, and Culture of the American West

Richard L. Knight

Ranching, as both an occupation and a process, is a vital element of the changing American West—ecologically, economically, and culturally. In a landscape that is increasingly more fragmented, ranching maintains connections between public and private lands, as well as between rural and urban communities. If ranching declines, so too will the health of human and natural communities in our region.

Ecologically, ranching keeps lands open and stewarded, keeps human population densities low, and safeguards against residential sprawl. Economically, ranching provides homegrown food and supports a fiscally responsible economy. Culturally, ranching is a heritage dating back over four hundred years, one of the oldest land uses that Euro-Americans have given the New World.

A natural alliance exists between urban consumers of food and open space, and rural producers of food and open space. This rural-urban partnership is essential to a healthy West, as is a strong public-private land connection. As these relationships deepen, so too will the health of the human and natural communities of this region. To best support this alliance, there needs to be a region-wide understanding of private and public land use as it pertains to ranching.

An honest appraisal of ranching as a land use in the New West first requires acknowledgement of the current "highest and best uses" of our land, both private and public. Based on monetary returns, economists have decreed that exurban development and outdoor recreation are the highest and best uses of the private and public lands, respectively, in today's West. In recent decades, these land uses have replaced livestock grazing as the principle use of the West, particularly the arid West.

Some people might think it is a far stretch to connect livestock grazing on private and public lands with exurban development and outdoor recreation on these lands, but I see it differently. The protection of open space, food production, ecosystem services, and of the aesthetics of rural areas runs right through ranching. We have arrived at a point in our history where conversations about western land and land health, grazing, ranchettes, and recreation are entwined and cannot be separated. They must be dealt with simultaneously when discussing the future of our Next West. Importantly, these discussions need to include more than just ecology—they need to address the economic and cultural aspects of these land uses as well.

Healthy Landscapes

Although land ownership in the West is blended, half public and half private, the division is not equal. The private lands are the best watered, occur at lower elevations, and contain the richest soils. As a result of the way the West was settled in the 1800s, federal lands today are largely "rock and ice" and "desert and thorn." The implications to biodiversity of this historical truth are critical. The private lands are disproportionately important to the maintenance of our region's natural heritage because they are more biologically productive. Although no one has calculated the ratio, private lands may be an order of magnitude more important to the maintenance of the region's biodiversity than are the public lands. Truthfully, however, species of conservation concern could no more survive on only the private lands of the West than they could on just the public lands. As the public and private lands are intertwined across the West, so too is the fate of our region's natural heritage.

Ranching, because it encompasses large amounts of land with few humans, and because livestock thrive on native vegetation, has been found to support biodiversity of

conservation value. The alternative uses of private and public lands do not tread so lightly. Outdoor recreation is second only to invasive species in causing decline of federally threatened and endangered species on public lands. On all lands, both private and public, sprawl is the number two cause for species imperilment, and recreation the fourth leading cause.

Whereas ranching is synonymous with few humans, roads, and houses, exurban development brings year-round activities and elevated human densities that fragment lands with roads and houses. For example, when ranches in Larimer County, Colorado, were subdivided, there was an almost ten-fold increase in roads and houses, which perforated and dissected the previously intact wildlands.

This observation led us to wonder how biodiversity, from songbirds to carnivores to plants, differed across the principle land types of today's rural West. Accordingly, we examined these species on a landscape that was part working ranches, part exurban developments, and part protected areas where livestock were prohibited.

We found that the ranch lands and protected areas supported birds and carnivores of conservation interest while the exurban developments supported pretty much the same songbird and carnivore community found in suburbs (imagine cats, dogs, and robins). The plant story was a little different. Both the protected areas and the exurban developments were far more weedy than the ranch lands. Stewardship, the judicious use of herbicides and livestock, and a discerning eye were the differences here. Ranchers apparently are doing what Aldo Leopold suggested when he wrote, *"The central thesis of game management is this: game can be restored by the creative use of the same tools which have heretofore destroyed it—axe, plow, cow, fire, and gun."*

The result of the biological changes associated with the conversion of ranchlands to ranchettes will be an altered natural heritage. In the years to come, as the West gradually transforms itself from rural ranches with low human densities to increasingly sprawl-riddled landscapes with more people, more dogs and cats, more cars and fences, and more night lights perforating the once-black night sky, the rich natural diversity that once characterized the region will be altered forever. We will have more generalist species—species that thrive in association with humans—and fewer specialist species—those whose evolutionary histories failed to prepare them for elevated human densities and our advanced technology. Rather than lark buntings and bobcats, we will have starlings and striped skunks. Rather than rattlesnakes and warblers, we will have garter snakes and robins. Is that the West we want? It will be the West we get if we do not slow down and get to know the human and natural histories of our region better, and then act to conserve them.

Of course, it goes without saying that ranchers are not all alike. Ranching, done right, can coexist with healthy land or even restore land back to health. Done wrong, it can damage and destroy. For instance, in the early days of the "open range," cows, sheep, goats, horses, and mules flooded the Western rangelands. In the absence of government regulations, this early example of "free-market" capitalism did long-term damage to grasslands, shrublands, and wetlands. By the early 1900s, however, range science was developing and land-management agencies were actively intervening to encourage grazing on a more sustainable basis.

Today, our understanding of rangelands and livestock grazing is beginning to mature. We've largely dislodged our previous tidy stereotype: livestock–overgrazing–desertification, followed by removing livestock–rest–recovery. Instead, conservationists are beginning to accept that nature is more complex than that. Ecologists are learning about the interrelatedness of climate, fire, and grazing. Grasses and shrubs co-evolved with herbivores, species that grazed and browsed their new growth. Whether by mastodons and sloths, or bison and pronghorn, or grasshoppers and rodents, grass and shrubs need the stimulating disturbance brought about by large, blunt-ended incisors clipping their above-ground biomass. Furthermore, efficient nutrient cycling is facilitated by the animal's dung and urine being incorporated into the soil by hoof action.

The West has long been defined by large populations of herbivores roaming its grasslands. Today the mastodons are gone, and there are fewer bison and pronghorn than once occurred. There are cattle and sheep, though not as many as we saw in the last century. But, we have learned that grazing by livestock, when appropriately done, contributes to the necessary disturbance that rangelands require. Perhaps we have come to the point where we measure land health premised on disturbance rather than just rest, and realize there is no "balance of nature," but instead a "flux of nature." Getting the disturbance patterns right is the challenge.

A word of caution regarding all of these findings. The West is not one place, but many places that grade into each other. They have different biological histories—and different ecological structures and functions—upon which cultural histories and land uses have been superimposed. Slope matters, as does elevation, aspect, and local rainfall. On a longer view, so does the post-Pleistocene environment. The implication is that some places may be more compatible with grazing by large, social, domestic ungulates than others.

Perhaps we should be humbled by all of this uncertainty. After all, it was a similar appreciation of the nonlinear behavior of arid Southwestern ecosystems that prompted Aldo Leopold to consider the need for a land ethic. To heal unhealthy lands we should seek counsel not only from ecologists but also from those whose connections to the land are long and deep. Perhaps these new vibrant partnerships between ecologists, conservationists, and ranchers have something important to teach us about how ecosystems function, and about how humans and land can get along better together.

Sustainable Economies

During a time when America's red ink is swelling large enough to swamp the world's biggest economy, it is encouraging to realize that ranching tends to be fiscally responsible. Study after study have all reported the same finding: Property taxes from rural residential developments fall short in paying the costs of county governments and school districts, whereas taxes from farm and ranch lands allow counties and schools to remain in the black. In Colorado, for example, for each dollar of property taxes collected from low-density rural sprawl, counties and school districts have to produce $2.15 of goods and services—which means they have to anti up $1.15 to meet costs. For farms and ranch land, however, counties and school districts show a surplus, having to only produce $0.35 of goods and services for every dollar of property taxes. In other words, Colorado ranchers are subsidizing the sprawling residential developments covering hillsides faster than Herefords can exit.

What about subsidized grazing on our nation's public lands? Ranchers are accused of feeding at the public trough. Wait a minute; what land use is not subsidized on our public lands? Indeed, outdoor recreation, our "highest and best use," is the most heavily tax-supported public-land use. This is appropriate considering that all of us, ranchers included, recreate on public lands.

Importantly, however, the American public benefits from allowing ranchers to

GREAT PLAINS

EAGLE ROCK RANCH

- *Weld County*
- *Owners: Cox Ranches, LLC; Mark and Emily Cox; Gordon and Beverly Black*
- *First ranched in the early 1870s*
- *Conservation easements held by The Nature Conservancy and Legacy Land Trust*

Tucked up against the Colorado/Wyoming border twenty miles east of I-25, Eagle Rock Ranch lies in and around the dramatic Chalk Bluffs. Northeastern Colorado was under a vast sea 80 million years ago. The sea drained, and streams crossing the area deposited sediments that hardened into today's sandstones and siltstones.

Benjamin Franklin Ketcham, great-grandfather of Beverly Cox Black, homesteaded the ranch in the early 1870s. Living in Wyoming at the time, he found the land while searching for lost cattle in a blizzard. The current larger ranch was assembled by Mary and Emily Cox in the 1940s. Prominent on the ranch is Squaw Peak, named for an incident involving one Elbridge Gerry. Gerry had recently traded for two Lakota Indian girls when he was attacked by Cheyenne, Crow, and Arapahoe Indians near the peak. He and the "squaws" escaped to safety by climbing the peak and pulling up the willow ladders they used to do so behind them.

Deer at sunrise

Ranch headquarters

SWEDLUND RANCH

- *Logan County*
- *Owners: Todd, Nichole, Kent, and Peggy Swedlund*
- *Ranched by the Swedlund family since 1948*
- *Conservation easement held by Colorado Open Lands*

Located eighteen miles north of the town of Sterling, Swedlund Ranch lies on a peninsula that stretches out into the heart of North Sterling Reservoir. Steep cliffs and rocky outcrops define the edges of the ranch, and cottonwood trees line the draws down to the reservoir. Short grass prairie occupies the bulk of the ranch's highlands. Mule deer, pronghorn, red fox, and a complete assortment of eagles and other raptors dominate land and sky. Native American teepee rings are easy to find; also of note is the "rock house," built in 1910—the same year as North Sterling Reservoir's dam—an old cabin made of native moss sandstone that is still used for storage and to house guests.

Moonrise

WRAY RANCH COMPANY

- *Yuma County*
- *Owners: Rex and Jody Buck*
- *Ranched by the Buck family since 1990*

Located ten miles east of the town of Wray, the north half of Wray Ranch lies among sand hills that are remnants of sand dunes formed ten thousand years ago at the end of the last ice age. The south half lies in remarkable spring-fed hills that nurture wetlands and irrigation reservoirs. Cottonwood trees decorate the riparian areas. Thomas Ashton first ranched the area in the 19th century, with registered shorthorn cattle and Percheron draft horses. Today, the Buck family runs a hay and cow/calf operation. Rex and Jody Buck each descend from fifth-generation Colorado ranch families from Kit Carson County.

G R E A T   P L A I N S

HILLSIDE RANCH

- *Yuma County*
- *Owners: Leta Smith, Larry and Kathy Smith*
- *First ranched by the Smith family in 1910*

Located along the banks of the Arikaree River, Hillside Ranch is characterized by magnificent cottonwood trees, pastures, and hay meadows. The Arikaree only runs intermittently, but has been known to flood violently when summer thunderstorms dump massive amounts of rain into its watershed to the west. Such a flood occurred in 1935, depositing large quantities of mud containing dormant seeds in the flood plain. The result is the forests that we see there today. The ranch is just upstream from Beecher Island Battleground Historic Site, where Cheyenne Indians and the U.S. Army had a major confrontation in 1868.

Ray Smith settled Hillside Ranch in 1910. His son Clare and his son's wife, Leta, were the second ranching generation, and today Ray's grandson Larry and his wife, Kathy, manage the operation. Their beautiful red barn was built prior to 1900 and is still functional.

Wild turkey

BRETT GRAY RANCH

- *Lincoln County*
- *Owner: The Nature Conservancy*
- *First ranched by Charles Thurlow around 1872*
- *Conservation easement held by The Nature Conservancy*

Fifty miles due east of Colorado Springs, Steels Fork Creek drains northwest to southeast for about forty miles. The ranch is located near the middle of the creek and contains lowland riparian pastures irrigated by it, as well as drier uplands. A few reservoirs interrupt the creek. The lowlands are incredibly fertile, supporting lush grass, shrubs, and cottonwood groves. It is truly an oasis in the midst of dryland prairie.

Charles Thurlow, originally from Massachusetts, settled the area around 1872. The land passed through several owners before 2007, when The Nature Conservancy acquired 23,300 acres of the heart of the ranch. Ownership of this acreage is set to be transferred to the Colorado State Land Board to create a 49,000-acre working cattle ranch in conjunction with the Board's adjacent holdings. The ranch will be managed by a ranching family. The protection of the ranch ensures preservation of the rural character of Lincoln County. Over two hundred playa lakes, home to the Plains Leopard Frog as well as migratory birds, especially waterfowl, will be conserved. The ranch's streams are excellent habitat for the tiny Arkansas Darter fish, a remnant of the Ice Age.

Common nighthawk

FOUNTAIN CREEK RANCH

- *El Paso County*
- *Owners: Ferris, Jay, and Harah Frost, Wallis Frost Mineo*
- *First ranched in 1864. Acquired by the Frost family in 1958.*
- *Conservation easement held by Colorado Open Lands*

Fountain Creek Ranch lies twenty miles south of downtown Colorado Springs. It is remarkable for its views of Pikes Peak to the northwest, and the serpentine and sandy Fountain Creek running through it. Wetlands along the creek are home to myriad waterfowl. The lower part of the ranch is irrigated for hay and alfalfa production.

It is said that in the mid-19th century, the Ute Indians would descend from their mountain homes in the spring and pass through what is today the Fountain Creek Ranch on their way to buffalo hunting grounds on the plains. At that time the Holmes family ranched much of the area. Old man Holmes would loan the Utes a rifle for the purpose of shooting buffalo.

Pikes Peak

JE CANYON RANCH

- *Las Animas County*
- *Owner: Jerry Wenger*
- *First ranched in the 19th century*
- *Conservation easement held by Colorado Cattlemen's Agricultural Land Trust*

The Purgatoire River drains the east side of the Culebra subrange of the Sangre de Cristo Mountains. It passes through the town of Trinidad, then carves a spectacular sandstone canyon, called Pinon Canyon or Purgatory Canyon, on its way to meet the Arkansas River near Las Animas. At one point it runs one thousand feet below the surface of Colorado's southeastern plains. On the north side of Pinon Canyon lies the controversial Pinon Canyon Military Reservation; on the south side are several magnificent ranches, including the contiguous JE Canyon and Beatty Canyon ranches. In the 19th century the Lopez and Gutierrez families ranched a portion of what is today JE Canyon Ranch. The Wenger family acquired and consolidated two large ranches spanning this portion of the Purgatoire in the 1980s to create the current ranch. Cattle graze both the canyon and the dryland prairie above it.

Purgatoire River

BEATTY CANYON RANCH

- *Las Animas County*
- *Owners: Steve, Joy, and Betty Wooten*
- *First ranched in 1864*
- *Conservation easement held by Colorado Open Lands*

Beatty Canyon Ranch on the Purgatoire River lies immediately downstream from JE Canyon Ranch. Therefore, many miles of what is also called Pinon Canyon is protected forever from development. The two ranches have similar geographical characteristics, and there is much natural and human history in both places. Hundreds of dinosaurs inhabited the area 150 million years ago, and over 1,300 dinosaur footprints are exposed in the canyon bottom near the ranches. The Santa Fe Trail, which served as a trade route between Missouri and the Mexican frontiers from 1821 to 1880, ran along the rim of the canyon. Indian rock art and other antiquities abound.

The Cordova family homesteaded Beatty Canyon in 1864. Their land was later acquired by Joseph "Papa Joe" Doherty, who immigrated from Ireland in 1879 at age 18, looking to make his way in the West. Papa Joe acquired much more land beyond the Cordova's holdings, including what is today's Beatty Canyon Ranch. His holdings were eventually passed down to his grandchildren, John and Betty Doherty. Betty's son Steve Wooten and his wife Joy run the Beatty Canyon Ranch. John and his son Joe run the Red Rocks Ranch next door.

Doherty passed a lot of this knowledge along to his nephew. Enough that when the ranch property was divided in 1989, Steve was ready to run the 27,000-acre Beatty Canyon Ranch while Doherty and his son Joe focused on their other 50,000 acres, made up of the Red Rocks Ranch and other properties. From his uncle, Wooten inherited a reverence for the land—and a disdain for invaders. Just as Doherty can't keep the indignation out of his voice when he talks about the way the profusion of red cedar trees is encroaching on his precious grasses ("These cedar trees, if the water's available, man, they'll take forty gallons a day out of the soil, so your springs dry up. Plus, if that tree wasn't there, you'd have that much more grass"), Wooten can quickly go from amiable to angry at the sight of tamarisk.

"The tamarisk is an ornamental that was introduced into this country—and what a mistake that was," he says, voice gathering itself into a growl. "It has a delicate purple blossom on it, but otherwise it's a disgusting plant. It drinks up a lot of water, then turns the soil alkali. That's the way it protects itself, it effectively kills off other plants by depositing salts back in the soil."

Cedar. Tamarisk. Whatever hurts the grasses that feed the cattle, whatever hurts the grasses that renew themselves and are part of God's handiwork, creates a boiling in the blood of the uncle and the nephew. Both love it, want to protect it. Both want to pass it on. Because, really, what else is there?

You spend your life ranching, chances are, says Steve, "You'll spend your life cash poor." Yeah, sure, you need to make a profit, but the question is how big a profit do you need? You don't work to protect your money, you work to protect your land. If you're John Doherty, you work for Joe, just like Joe works for Sean and Matt, his two boys. If you're Joy and Steve, you work for Niki and Arin, just like they'll work for Avery and Bray and whoever else might happen to come along. You work like this, says Steve, because "Success is passing this land down to the next generation, and the one after that."

Over the chugging drone of a rusty-blue pickup truck that has more miles than anybody knows because after 140,000 the odometer isn't worth spit, Doherty is talking about his land and why "It's the best place in the world to raise a family." As the words stroll out of his mouth, you can't help but be reminded a little of Henry Fonda, the actor who brought quiet honesty and earnestness to every role. There just doesn't seem to be any room at all for doubt to slip in when Doherty is talking. So you listen. You listen when he talks about the ranch being a place where "your family lives together and works together." Where everybody has chores early on, even "those two little guys," his two grandsons, eight-year-old Sean and four-year-old Matt. You listen when he sits back in that pickup that's chugging so loud sometimes you can barely hear him, and you strain to catch the words, "This kind of environment to raise a man in is just, well, it's just *uncommon*."

But sure not easy. Hell yes, it's tough. Hell yes, you "need to be optimistic because this is a business of next-year type deals. You know, if you're having a bad year, you have to think, 'Next year'll be better.' If you're in a drought, you have to think, 'Well, it's one day closer to rain.' "

Steve Wooten knows what his uncle is talking about. He knows about droughts, like the one in 2002 that forced him to sell off 225 head—about half his herd—and take the rest to Kansas to graze. He knows about winter storms, like the one in 2007 that dropped forty inches of snow on his land and, just for good measure, conjured up winds that sculpted the snow into twelve-foot drifts. He knows that you just have to do what you have to do. During that same winter storm he went out and leased a D-9 Caterpillar for thirty days—at $500 a day—and plowed his own land. And then his neighbors' land. Opened up county roads. Opened up fissures in the snow to bring food to stranded cattle. And when neighbors wanted to pay him for his time and trouble, he took a pass because that's what neighbors do for neighbors.

But fifteen grand is fifteen grand, and so even while working in a business of next-year type deals, sometimes you have to look at different paths to next year, paths that might be as uneven and winding as the seven miles of dirt road that form the "driveway" connecting the Wooten ranch to County Road 177.9.

For Joy and Steve, the prospect of selling off part of the land to developers eager to turn it into vacation-home parcels wasn't a possibility. Not when there was the next generation waiting to work the land and love it. Dissect the canyon? Hell no. And to make sure *that* would never happen, they obtained a conservation easement, something to protect the heritage they wanted to leave behind. Still, the dicey nature of the cattle business meant, as Steve knew, "it's difficult not to have alternate sources of revenue to break even." That's why he and Joy are trying to find ways to host artists seminars, guided hunting parties, maybe just-plain vacationers to the ranch. That's why "I've been working on building a lodge out next to the barn for three years."

"Four years, honey," interrupts Joy. "You've been at it four years."

"You had to rub it in," replies her husband, laughing and smiling at the teasing like it's something fresh and new instead of a thirty-year pattern of puppy love and affection.

More people coming to visit and appreciate their land makes the Wootens happy. Because sometimes when strangers show up, they aren't the only ones who feel a connection. The strangers aren't the only ones who learn things about the land. Sometimes it's the Wootens who learn. Like the time they discovered their land could heal.

Earl and Linda were in their seventies. Linda was dying of cancer. Earl knew that. He also knew that, more than anything, she wanted to be able to get back on a horse and just . . . ride. The way she had when she was young. He called the Wootens and asked if they could help.

They found a gentle horse for Linda. They went riding. Easy stuff. Toward Eagle Rock. Into the embrace of the sandstone and the grasses and the cholla cactus that always look so beautiful in late spring when they explode with purple flowers. Into land where you can see what's left of the Cordova homestead, which isn't much, maybe a building and whatever pieces of adobe walls haven't melted into the ground. The Cordovas built their home in 1864; they were the first to ranch the land, the first to coax it into production. They were the ones who built the cemetery up the rise from their compound, the cemetery where bleached headstones still whisper to you that *Evanjelina Cordova* died *Jun 10 1906* when she was just *8 meses 17 dias*.

Across the canyon floor from the cemetery are immense sandstone outcroppings. Inside one, near the shaded coolness of a ten-foot-deep pool of melted snow and rainwater, there are names carved into the rock. Names like *Magdalena Cordova* and *Ben Pruet 1860* and even *Kit Carson*, who was likely going to homestead here but never made it back. Someday, there may be another name to join the roster. Someday Steve Wooten may carve his name on the walls, although "Not in this area. Maybe over there. Just to carry on the tradition of the people who came here before us. You know, to say, 'I was here. I passed through here, too. I appreciated it and enjoyed it.'"

It was through this land that Earl and Linda rode with the Wootens, land that Earl called "Shangri-La." Land that he knew had touched and restored part of his wife and had been a beautiful gift to her before the cancer finally took her.

By the time they finish telling the story, Steve and Joy are both crying. It's hard to say whether they are sad or happy about the memory. Maybe both. Healing can be funny that way.

Morning is on the verge of turning into afternoon and the branding is going like gangbusters. There are so many helpers in the corral that there are two ropers and three teams flanking calves. Dowell and Jan Mapes are moving between the animals, vaccinating them. John and Jerry Winford are moving between the male animals, turning bulls into steers with surgical precision and efficiency, carrying away the excised testicles in their bloody hands and dropping them into the plastic bucket that holds a growing feast for Doherty's dog. Men with freshly turned irons move between the animals, kicking up fountains of dust, heeding the call of "Need a bar…need a triangle…need an L," careful to put the right amount of pressure with the right amount of heat so the brand will take.

In the middle of this bigger ballet, ranchers whirling around him, the propane oven whooshing like a steady gust of wind, sits eight-year-old Sean Doherty, not one whit distracted. He mans his post, diligently putting numbered plastic identification tags in a device that will affix the tags to the calves' ears. His face is serious. He's not saying much. Two hours of this and he hardly ever takes a break. He's got his job and he knows it. He's growing up just the way his grandfather and his father want. The ranching way.

Then, barely ninety minutes and 121 calves later, it's time to stop and shift gears. At least temporarily.

"This was good," says Winford. "Sometimes things can get a little wild. This was a good morning. Nobody got bucked off or run over, and that's always a relief. Sometimes, one of them calves comes through and it'll knock over the whole thing. Branding irons, everything. Not today. Yeah, this was a good morning."

As efficiently as they branded, the helpers clean up, quickly packing boxes of vaccinations and assorted tools onto a rusted-blue pickup. Of course, now the incentive is even greater. Down at the Doherty house, in Chacuaco Creek Canyon, food is being laid on the table by Carolyn, John's wife; Betty, John's sister and Steve's mother; and Lisa, Joe's wife. It's all a little bit of eating heaven—roast beef, potatoes, corn casserole, coleslaw, Jell-O, rolls, three different kinds of pie. Not a piece of tofu in sight.

Grace is said and food is swooped up. There's a lot of eating, a lot of talking, a lot of laughter. And no small amount of gratitude from John Doherty.

"It's been a good day, a pretty day, good friends," he says, looking out at a scene he's seen more times than he can count but never gets tired of because how can you ever become jaded about good people rallying to your side when you need a little help? He looks some more and in that Henry Fonda-earnest voice says softly, "Everybody's so nice and slowed down now, it'd be nice if we were through and everybody could just relax and spend time with us as long as they wanted."

Then he walks over and starts putting on his boots because he knows they're not through. There's seventy more calves in a pasture up those canyon walls and even a nincompoop greenhorn can figure out that nobody's going to be relaxing and sitting around passing the time. Hell, there's work left, calves to be branded, fellowshipping to be enjoyed for sure, but still work. And it needs to be done before night comes along and chases the color out of the sky and all that's left are the stars shining down on God's handiwork.

PATTERSON RANCH

- *Las Animas County*
- *Owners: Bob, Bunny, R.C., and Joanna Patterson; Kandi and Bruce Nittler*
- *First ranched by the Patterson family in 1967*
- *Conservation easement held by Colorado Cattlemen's Agricultural Land Trust*

Though the Pattersons ranch some of the flatlands south of the town of Kim, it's the Mesa de Maya that defines their ranch. At fifty miles long and up to six miles wide, it's one of the largest free-standing mesas anywhere. Its maximum elevation of 6,800 feet makes it the most conspicuous thing rising above the southeastern plains east of Trinidad. Travelers can only imagine what's on top because it's pretty much all private land—much of it part of Patterson Ranch. Ponderosa pines occupy the highest reaches with junipers everywhere below.

Good grass grows in the fertile soils composed from the volcanic lava flows that created the mesa. Cattle and sheep have been grazing here since the late 19th century. The difficult access to the top made it the perfect hiding place for outlaws, and rumor has it that Billy the Kid and Pretty Boy Floyd enjoyed respite from the law there.

But it's not just the vast breadth of the Mesa that detonates so much joy and wonder inside Bob. Part of the land's magic is the living, breathing treasure it contains. See, "There's quite a lot of black bears, antelope, mule deer, elk, sheep, mountain lions, bobcats, turkeys. A lot of foxes up there. Antelopes everywhere. Coyotes, too. Only since they raised such a fuss about killing 'em, they don't let you sell the hides. Kids used to hunt 'em—as a matter of fact, I know some kids used to sell enough coyote hides to buy a pickup."

Bob himself hasn't hunted in "Oh, fifteen or twenty years." As he got older, he decided when it came to deer and elk, "I'd rather watch 'em than shoot 'em." Not that he has a problem with hunting. Hell, he leases part of his land to a guide who arranges trips for people during the hunting season. He'd just rather look now. Just like he'd rather be seeing the Mesa from behind a pair of reins instead of a steering wheel.

"What I really, really enjoy is horseback," he says. "I can sit on a horse all day long. You don't get to do that near enough today with all the hustle-bustle, all the things you got to take care of. You got to be on a pickup or a tractor or go to some meeting at the bank. That's why horses aren't nearly as good today as they once was." The truck strains up a steep grade and around a curve. In the distance, Bob points out a "big ol' little mesa over there. See it? That ain't part of Mesa de Maya, though."

He gets out of the pickup to open a barbed-wire gate and checks a homemade rain gauge. Nearly three-quarters of an inch of spring rain has fallen, a providential rain, especially when the past two months have been dry as bones. The wind is starting to kick up. High above, it shapes and reshapes the rolls of cumulus, almost like it was working in some kind of cloud laboratory. Everything is cloud and sky and prairie grasses and, by the way, if you want to know what it is that's growing all in front of you, well, step back, shut your mouth, and get ready for a graduate course in grass. He may have been a "less than average student" in school, but outside, up on the Mesa, Bob Patterson is pure professor.

"Okay, now this right here is called western wheat—or vega. The shorter grass is gramma, blue gramma, and there's a little buffalo mixed in there. And there's some thread-and-needle—that's the bushy stuff over there. And then there's that little fuzzy grass, see it? I call it 'cheat grass.' It's got a name that's this long, but it don't do much for cattle.

"That mushy grass is snakeweed. It's a real bitter taste so the cows don't hardly ever eat it, but if a cow does, she'll abort a calf. Now see that white stuff? That's locoweed. If the cows get on that, they'll abort calves. If you got plenty of green grass, they won't bother with it; it's like the snakeweed, real bitter tasting. But if they come up on some and eat it, it's like marijuana. Sometimes they'll come out and bite ya. Or they'll stand with their heads just swaying like this."

The wind riffles the grasses, which still haven't greened up yet. Soon they will. In fact, "In another month, this whole place'll be filled with flowers."

"What kind of flowers?" you ask.

"Well, there's paintbrushes and, those red and yellow flowers. Aw, I can't think of what they're called. Bunny's the flower person. I just call 'em 'wild tulips.' That gets her mad at me. C'mon, let's get out here, I wanna show you something."

The something turns out to be the alleged hideout of famed 1930s bank robber Pretty Boy Floyd. Legend has it that when things got a little too hot in Chicago, Floyd would head for the Mesa. Sanctuary from the cops. A place to clear his head. Kick back. Maybe do some hunting. After all, he was a Georgia boy who'd done a lot of growing up in Oklahoma.

"This here's his bungalow," laughs Patterson.

You stand in front of a twelve-by-twelve log cabin. Or what's left of it—which isn't much. Four walls, each down to about four feet high. No roof. Part of an old rusted stove in front. Of course, no one knows for sure if this was Floyd's hideaway, admits Bob. "The two guys that told me about that [Pretty Boy] are both dead, so I don't know. But where there's smoke, there's usually a little fire."

Besides, it makes sense that Pretty Boy would have come here. The remoteness of the Mesa, the less-than hospitable road to its top, would have made it a good hideout. Just like it would have made a good hideout for the likes of Billy the Kid and Black Jack Ketchum, two other outlaws who, if you're inclined to believe local lore, were no strangers to Mesa de Maya. Hey, it's not coincidence that there's a place on the eastern edge called "Robber's Roost."

Maybe the likes of Pretty Boy and Billy the Kid are more fodder for legend than history. But that doesn't mean bandits still don't lurk near by the Mesa. Bob Patterson knows they do. He's seen them, spoken to them—modern-day robbers trying to steal the land and corrupt it. And that's no fiction.

Bunny knew the perfect way to get a rise out of Bob. They'd be riding across the Mesa on one of those days where everything looked so wild and beautiful you could swear you'd stumbled into Eden. And whenever they'd stop to look at forever, she'd say, "Now, wouldn't that spot make a nice site for a cabin?" And, just like that, Bob would get ticked off because if there was one thing he hated to think about it was a cabin, or one of the "ranchettes" or phony-baloney lodges, being built in a meadow, being built on sacred ground.

That's not what this land was meant for. No, the land was meant to accept the sun and drink up the rain and grow the grass that fed the cattle that helped feed the world. Bunny, being a teacher, knew all the statistics by heart. "There's less than two percent of us raising food in our country; less than two percent of us to feed the world. Did you know that each rancher feeds 187 people each day? That's three meals! Now you multiply that three meals a day by 365 days and we have a huge responsibility. We can live a long time without a lot of things, but food isn't one of them."

She and Bob knew that some environmentalists didn't understand what ranchers do. She knew that when her husband said, "A lot of them don't know squat about the environment" he was dead on. She knew that to keep a short-grass ecosystem healthy and flourishing, it must be grazed. Not indiscriminately, but carefully, intelligently. With an eye toward making sure there was no overgrazing so the land would perpetuate its gifts. She knew that "concrete and asphalt are the last crops" and when you start building rustic show homes and breaking up the Mesa into—god help her—plots, something's gotta give and that something is the land's soul. She knew that "Ranchers are the original environmentalists," because to them the land isn't something you vacation on or smother with concrete. That's why her voice would sink to a whisper that had ten times the passion of any shout whenever she put her face a little closer to yours and said, "Our land is our life."

That's how Bunny and Bob felt. Their children, too. Why, after his rodeo days were over, hadn't R.C. come straight back to his roots and started ranching? Hadn't he done that even though he knew "Most business people would never invest as much money as ranchers do for so little profit"? Hadn't he and his wife Joanna patented a feed-processing technology that could cut feed costs in half and started marketing it on the internet because "If you want to live on the land, you have to find other sources of income"?

You work this hard and sweat this much, you're not about to sit by while cabins start popping up like so much snakeweed. And that's why the Pattersons—and Kandi and her husband Bruce Nittler—went out and sold easements on their land. They may have been running separate ranches, but their vision was united: keep the land whole and productive. Easements seemed the logical thing to do, the right thing to do—but that's not how some people saw it. Instead, according to R.C., "They thought this was the devil at work." Or something even worse.

More than a few ranchers heard the word "easement" and the first thing that popped into their minds was government. Big government. Big Brother. Why, to hear what some people were saying, what Bob and Bunny and their kin had done was open the door for government to take away everyone's land, more than likely sooner than later.

No amount of explaining satisfied some of them. Old friends became ex-friends and if you asked Bob about that he'd say, "Maybe they weren't friends after all." And if you asked him if it hurt, Mr. Pollyanna would fold his arms, shrug and say, "It don't make me no difference." And then, just like always, he'd move on, figuring things would get better.

Besides, he knew what they were doing was right. He knew they were protecting the land. Not just for him and Bunny. Not just for R.C. and Kisti and Kandi. But for their seven grandchildren and their grandchildren's children. It was like Bunny said, "Our forefathers went through a lot to save the land. It's up to us to save the land for what it was meant to be used for—ranching."

In 2001, when the easement paperwork was all signed and official and a good portion of 19,000 acres of sacred Patterson and Nittler ranchland was protected, Bob and Bunny gathered friends and family up on the Mesa. Then Bob got up and said a few words about how glad he was that the land would be preserved after he was gone, and among those words were, "If I'm up in heaven, I want to be able to look down on cattle. And, well, if I'm below, I want to be looking up at cattle."

Which, when you think about it, makes one fine big ol' little epitaph for a rancher.

"That's New Mexico, and over there is Oklahoma, and right there is Trinidad, Colorado. And you see over there—that little line of trees? That's the west pasture. See this highest peak way down there? That's El Capitan and that's a volcano. It's in New Mexico. See over there? That's an elk ranch. Guys from New York come out and pay ten thousand dollars and they give 'em cake to give to the elk and then they shoot 'em. What's the fun in that?"

Bob Patterson is shaking his head, unable to figure why anyone would spend all that money to pretend he was hunting. He knows a lot of those people who'd spend all that money live in big cities and drive fancy cars that don't have 200,000 miles on them and "probably consider us to be 'hayseeds.' But that's just their perceptions." Hell, most of them don't know gramma from grandma, cowhide from cow shit. "They're the ones if you ask 'em where milk comes from they'll say, 'The store.' "

The wind has stopped dead and in the quiet—a good, pure quiet—there's time to just be. Maybe Bob is thinking of his father, thinking of the way the land got into his father's blood, the same way it got into his. Maybe he's remembering how, when his father's knees and eyes went bad and he got too old to ride and work and be a rancher, he went to his mother and said, "Mom, dad ain't going to last long." And how he didn't.

But that was to be expected. Everybody's number comes up. Everything comes full circle and ends. But if you're strong and stubborn and lucky enough, maybe you get to pass on the only thing that lasts. The thing you worked hard to protect and preserve. Because up here on the Mesa, forever isn't just about distance—it's about a promise.

ROBERTS RANCH

- *Larimer County*
- *Owners: Catherine Roberts and the Roberts family*
- *First ranched by the Roberts family in 1874*
- *Conservation easement held by The Nature Conservancy*

The Roberts Ranch lies at the heart of the Laramie Foothills and has been identified by scientists at The Nature Conservancy and elsewhere as a priority landscape due to its biological significance. Protected in this easement are areas of historical interest, such as tepee rings, a portion of the Overland Trail, and at least one buffalo jump. It is also significant to wildlife because it is situated near over fourteen thousand acres of state and locally protected lands, thus increasing the area available to wildlife for migration, resting, and feeding.

The beginnings of the Roberts (Livermore) Ranch, which now encompasses 16,500 acres, date back 130 years. On July 3, 1874, the first members of the Roberts family arrived in what was then called Livermore Park to manage land and cattle for Greeley resident Russell Fisk. R. O. Roberts started buying his own cows and heifers, along with a team of horses, and made a "squatter's claim" along the North Fork of the Cache la Poudre River to start his ranch.

Nineteenth-century homestead

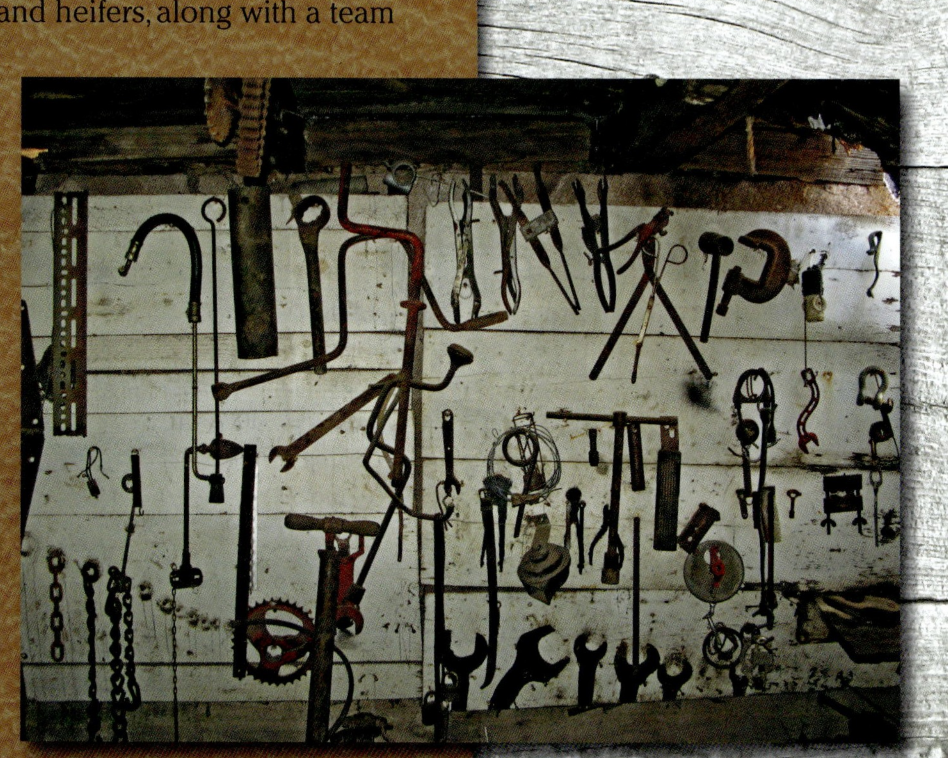

Views to the Medicine Bow Mountains

DEER VALLEY PARK ASSOCIATION

- *Park County*
- *Owners: Woodward and Dixon families*
- *Mined originally for copper in 1867, then ranched thereafter*

The ranch lies just a few miles east of the town of Bailey along U.S. Highway 285 and contains spectacular ponderosa pine woodlands, riparian habitat, and irrigated hay meadows. Pink granite outcrops that are part of the Pikes Peak formation dominate the edges of the ranch. Deer Valley Park Association was originally formed as a mining corporation. Today, it is one of the remaining cow/calf ranching operations close to Denver.

View looking south over U.S. Highway 285

Outhouse

DAKAN RANCH

- *Douglas County*
- *Owners: Catherine Sinclaire Brooks and Steven Brooks*
- *First ranched in 1870 by the Dakan family*
- *Conservation easement held by Colorado Cattlemen's Agricultural Land Trust*

Dakan Ranch is tucked into the foothills of the Rampart Range south of the town of Sedalia. It is part of the bucolic, if not Kentucky-like, series of farms and ranches on display along State Highway 105. West Plum Creek and the magnificent red sandstone outcroppings that ascend above the oak brush along the base of the range are the significant landmarks. These rocks are part of the same geologic formation found in Perry Park, Red Rocks Park, and others places up and down the Rampart Range.

William Allen Dakan and his wife Elizabeth homesteaded West Plum Creek valley along with fifteen or twenty other families. Ute Indians had a presence in the valley, and made life interesting for the families by begging for food and playing practical jokes—like jumping out of the bushes and scaring the children! Dakan descendents owned the ranch until 1952, when it was purchased by William W. Sinclaire.

Chris West poses on sandstone

BAR OPEN AK RANCH

- *Teller County*
- *Owners: Mark and Jim Johnson, Jim Hammond*
- *First ranched in 1928*
- *Conservation easement held by Colorado Cattlemen's Agricultural Land Trust*

Bar Open AK Ranch is a "cherry stem" of private land surrounded on three sides by Bureau of Land Management land south of Pikes Peak. East Beaver Creek bisects the ranch and irrigates beautiful meadows that define the ranch's bottom lands. The balance of the property consists of thick forests of ponderosa pine, spruce, fir, aspen, and scrub oak that cover steep hills above the creek. Most conspicuous are the eroded pink and orange granite spires and outcrops of the Pikes Peak massif, part of a one-billion-year-old batholith that extends all the way north to the drainages of the South Platte River.

Circus and mail aviator Arthur Hammond homesteaded the ranch in 1929. Today, grandson Jim Hammond and brothers Mark and Jim Johnson manage the ranch in its protected status. In 2002 the Johnson brothers entered into a partnership with the Rathke family, heirs to the legendary Rathke Elk Ranch founded in 1888, in order to expand protection of the East Beaver Valley to almost 2,100 acres.

The Johnson cabin

East Beaver Creek

The view north

Ponderosa pines

3R RANCH

- *Pueblo County*
- *Owners: Reeves and Betsy Brown*
- *First ranched in the late 1880s*
- *Conservation easement held by Colorado Open Lands*

3R Ranch lies twenty-five miles southwest of downtown Pueblo against the foothills of the Wet Mountains. It is remarkable for its vast irrigated hay meadows, views of Greenhorn Mountain, and the surprisingly rugged canyon of the St. Charles River. The canyon courses through the north end of the ranch and is an isolated ecosystem well removed from ranch activity. It is a retreat for raptors, deer, mountain lions, and many other creatures. Ute Indian teepee rings and other antiquities abound. One Annie Blake acquired the property as a land grant in 1868. Thereafter she sold an interest in it to Peter Dotson and legendary Texas rancher Charles Goodnight. The Brown family acquired the ranch in 1981. The name 3R dates to the ranch's origins and describes the branding of an R on the shoulder, ribs, and hip of the cattle.

Ranch headquarters

St. Charles River canyon

Hay meadows below the Wet Mountains

SAM CAPPS RANCH

- *Huerfano County*
- *Owners: Frank and Sue Menegatti*
- *First ranched in 1873*
- *Conservation easement held by Colorado Cattlemen's Agricultural Land Trust*

Sam Capps Ranch lies just east of the Spanish Peaks south of the town of Walsenburg. In fact, 12,683-foot East Spanish Peak is visible from every part of the ranch. This vast property is dominated by some of the most stately ponderosa pine trees in Colorado. Hundreds of elk live on and migrate through the ranch because of the lush pastures throughout, and many other Colorado creatures call it home at one time or another.

The ranch was homesteaded after the Civil War by the Capps family. It was lost in the Great Depression but reassembled thereafter with the determination of third generation family member Sam Capps and his father.

Elk herd

Black bear

Elk calf

Ponderosa pines

Golden banner wildflowers

Sunrise on East Spanish Peak

CANTRELL RANCH

- *Huerfano County*
- *Owners: Joan and Bruce Cantrell*
- *First ranched in the 1890s*
- *Conservation easement held by Colorado Open Lands*

Cantrell Ranch lies west of the Spanish Peaks, in and above the valley of the Cucharas River, south of the town of La Veta. Similar to Sam Capps Ranch, from which one sees East Spanish Peak, all of Cantrell ranch commands views of 13,626-foot West Spanish Peak. In addition, massive rock walls—volcanic dikes that radiate out from the two peaks—make this landscape unique in Colorado. It was first ranched for cattle by the Goemmer family. Brothers John and August also worked as blacksmiths on the Denver and Rio Grande Railroad that once crossed nearby La Veta Pass on its way west into the San Luis Valley. Today the Cantrell family raises horses on the ranch.

West Spanish Peak

FETCHER RANCH

- *Routt County*
- *Owners: Jay Fetcher and family*
- *First ranched in 1886*
- *Conservation easement held by Colorado Cattlemen's Agricultural Land Trust*

The Fetcher Ranch lies north of the town of Steamboat Springs along the banks of the Elk River. The Elk River, which meets the Yampa River near town, drains primarily from the Park Range and Mount Zirkel Wilderness to the east. These mountains dominate the views from most places on the ranch. Yampa Valley contains many ranches protected, like Fetcher's, by conservation easements, making it one of the most bucolic and least developed mountain valleys in Colorado. Vast hay meadows line the banks of the Elk River for thirty miles from Steamboat to Steamboat Lake State Park. Brothers John and Stanton Fetcher purchased the ranch in 1949. John's son Jay took over management of Fetcher ranch in 1968, while he was still in college.

Steamboat Lake and Park Range in the distance

Although things definitely improved, the enterprise was always a drain on John Fetcher's energy and finances. By 1969, when it was apparent the ski area needed more capital for a gondola and other improvements, he and his partners sold the renamed Steamboat Ski Resort to a Texas outfit. By then the monster was lumbering out of control. The ski resort drew people in. Land prices began to escalate. Developers licked their chops. Fancy houses, trophy homes for part-time residents, began to dot the hills. Slowly but steadily, ranching was on its way to becoming an afterthought in the valley's consciousness and economy. Recreation—winter skiing, summer boating and hiking and golf—that was the ticket. Families who had been raising cattle and growing hay for over a century were starting to be looked upon as anachronisms; quaint throwbacks to the past, definitely not part of the future.

By 1993, land that John and Stanton Fetcher had paid $70 an acre for in 1949 was going for $2,000 an acre—and climbing. More houses were going up. Golf courses, too. Something had to be done, and Jay Fetcher—who'd been running the ranch for a quarter-century—was about to do it.

That spring, he and five of his neighbors gathered at the Fetcher house to share fears about what was going on and find ways to stop it. The group knew that ranching wasn't only about profit. It was about lifestyle. About something in your heart that had to be felt to be understood; something that couldn't be communicated by words alone. Among the people who sat drinking coffee on his deck, feeling the afternoon breeze on her cheek, looking out at Sand Creek, watching chickens strut and cows amble, was Mary Mosher, an eighty-eight-year-old native of the valley. At one point during the discussion, Mosher clenched her fists so hard her hands turned white as she insisted, "I want this land to stay the way it is!"

Jay Fetcher was among the first to assure that would happen. With the unbridled support of his father and family members, he consigned all 1,250 acres of his family's Clark Ranch to a conservation easement with a group called American Farmland Trust, restricting it to agricultural use, protecting it from a developer's backhoe, ensuring no fancy homes would ever crowd out the cows or the hay. He also ensured the value of his property would plummet by more than half.

Others in that meeting followed suit—eventually, more than 7,500 acres of land in the Upper Elk River Valley would be protected. But Fetcher didn't stop there. Over the years, he began spreading the gospel of easements to other ranchers. Often, they were skeptical. As he quickly learned, "The two worst words in a rancher's vocabulary are 'conservation' and 'easement.'" He heard from land-rights groups who cautioned that "easements are just a government plot to take your land." He knew about the percolating suspicion that land trusts were run by conservationist zealots, environmental wack jobs, people who thought the best way to save the land was to leave it completely alone.

Sure, he understood the concern and the wariness. He knew that if there's one thing ranchers don't like it's being told what to do on—or with—their land. But, long before most, he also understood that maybe the only way to keep that *feeling* in a rancher's heart from going from flame to flicker, the only way to keep the land restricted to agricultural use—saved—so that it could be passed along to all those generations to come, was through easements. And he knew that instead of working with an out-of-state association like he had, ranchers in Colorado would be better served by something homegrown; they'd be more comfortable if they could ask questions and hear a familiar voice giving back answers. So he took his understated-but-persistent crusade into the maw of resistance: the Colorado Cattlemen's Association.

It wasn't a cakewalk, but in 1995 Fetcher and like-minded members of the association won approval to form the Colorado Cattlemen's Agricultural Land Trust (CCALT), the first U.S. land trust established by an organization of agricultural producers. Two years later, for his efforts in helping to create the locally run 501(c)(3), Fetcher was awarded the prestigious American Land Conservation Award, and an accompanying $50,000 grant. By 2008, the CCALT held 202 conservation easements that protected over 320,000 acres in thirty-five counties across Colorado. And of that total, more than 1,700 acres belonged to the Fetchers. Because in that same year, Jay decided that his family would also be better served by a homegrown organization with a familiar voice and transferred the original easement with American Farmland Trust to the CCALT.

Today, with all but 170 of the Fetcher Family's 1,900 acres under the aegis of conservation easements, Jay Fetcher is content. Sure, he knows that he will never realize the $15,000-an-acre price some of his land could likely fetch from a developer. But if he knows he'll never wrap his hands around that money, he also knows his fist will never be squeezed white in defiance or frustration. Instead, his hand will be free to tap his heart and feel the place where he stores his love for the land. It doesn't matter that he might wind up to be the last Fetcher to work the land. It doesn't matter that after he and Gael are gone, there's a good chance that none of his three daughters or grandchildren—or future grandchildren—will harvest a hay crop, or spend endless hours trying to fix fences that are annually broken by winter snows, or slog through a 2 a.m. storm to deliver a calf. It doesn't matter because when you make the decision to forgo big money in order to keep golf courses and manicured lawns off ranchland, "You don't do that for your kids. You have to do that for yourself." That's a choice you have to make on your own, the choice that Jay Fetcher made.

The wonder is that he ever got the chance.

He had grown up with a nanny. He had lived in Europe and learned to speak fluent French. He had been sitting in the Olympic Stadium in Berlin in 1936 when Jesse Owens won a gold medal and tore a large hole in Hitler's balloon of Aryan supremacy. Then John Root Fetcher, whose grandfather had been the architect of the 1893 Chicago World's Fair, whose lineage was more Brahmin than common, came back to America, attended Harvard, settled in Bryn Mawr, Pennsylvania, and started what seemed would likely be an inexorable rise to corporate and social success, Eastern style.

Except that somewhere along the way he and his wife decided something was wrong with the scheme. The need to join a better country club and diligently cultivate social status, the need to confront "forty stoplights on your way to work everyday"—it was just too much stress, too much bother, too much doing what they didn't want to do. John's brother, Stanton, felt the same way. The two men sat down, thought about their options, and figured, "Hell, anybody can be a rancher."

Which, of course, as they soon discovered, was completely nuts. But if the Fetcher Brothers were card-carrying members of the country-club elite, they didn't put on any airs when it came to hard work. John, in particular, even outworked the locals. A six-day workweek? No way he would acknowledge Sunday as a day of rest. Even Christmas was no excuse to avoid ranch tasks.

Another thing his neighbors noticed about John Root Fetcher was he had the same attitude as any good rancher when it came to production. You irrigate to grow grass. Then you sell some and feed the rest to your cattle. From the cattle you produce food. Ranchers make something, ranchers *produce*. There was nothing idle or pie-in-the-sky about a John Root Fetcher project. His ranch, the dams he'd build, that ski area—they served a purpose.

Still, if he won over the valley's ag people by his work ethic and his utter willingness to embrace the neighbor-helps-neighbor credo of ranchers, John Fetcher was still a bit of an odd duck. How could he be otherwise? He had come to a place where fifty years of residency still made you a newcomer. He and his brother had built a house with a basement and then watched as the basement promptly flooded, a minor catastrophe that could have been prevented if only they—like all their amused neighbors—had known a damn thing about the valley's water table. Hell, for a while everyone suspected he didn't even know which end of a cow got up first. And how come the Fetchers didn't eat their big meal of the day at lunch like all the other ranch families did? How come you never heard him or his kids calling a creek a crick?

But he had grit and smarts, and those neighbors who may have smirked when John Fetcher showed up were soon won over and became his friends. Those friends watched how the Fetcher children—Ned, Bill, and Jay, and then little Amie—were expected to handle their ranch chores. Of course, it didn't take long for anyone to figure out which one was the *real* rancher, which one loved the animals, which one could tap his heart and know that *feeling*. It wasn't Ned—he got shipped off to boarding school early on. It wasn't Bill—he joined the Navy as soon as he was old enough. No, it was Jay, small, quiet, introverted Jay, who flowed with the land. Years later, in fact, he would joke that, considering how little his father ever learned about animals, "I'm sort of a first-generation rancher." And who could doubt him? Why, hadn't he gone and gotten himself busted up just like a genuine ag man? Happened when he was no more than four years old; lost part of a finger to a farm machine. Yeah, that boy, he was the real deal.

The youngest son grew up on a regimen of work. He never had time for sports at school because there was always something to do on the ranch. Some kids played ball; he milked cows. Or did irrigation. Or drove a bulldozer. Or did haying. All that hard work toughened him, but it made him shy among his peers. He always felt like he was playing catch-up at school because it sure was hard to hone your social graces when just about all your spare time was spent conversing with cows. Of course, nurtured by his mother, he did have time to think, to grow philosophical. Not many ranchers would evolve into men who could quote Mother Teresa to strangers ("You can do no great things, only small things with great love") the way Jay could.

If the hard work was ceaseless, there were compensations for being the rancher heir-apparent. When Jay was thirteen, his father sent him up to work on the Hahn's Peak Ranch for the summer. There he fell under the sway of Ernie Murphy, a gentle, patient soul who was the ranch foreman. It was Murphy who taught Jay how to break a horse, how to read cattle, how to be a rancher. It was Murphy who also introduced the kid to an alien concept. The first Sunday he was up there, Jay asked Ernie, "What are we gonna do today?" "Nothing," replied Ernie. "It's a day off." *A day off—what's that?*

Jay spent two summers learning from Ernie, and when the old cowhand finally moved on, he spent three summers at the upper ranch all by himself, working cattle, fixing fences, doing the irrigating, learning self-sufficiency and independence. It's hard to say when, but somewhere along the line Jay figured out that he knew a whole lot more about ranching than his father. Eventually, his father figured that out, too.

John, Sr.'s obsession with building and running the ski area had taken its toll on the ranch. Even when Colorado State Parks got him involved in the construction of Steamboat Lake State Park—and paid him handsomely for his 900 acres of land that got swallowed up by the recreational facility—John Fetcher struggled as a rancher. The Parks payoff had basically saved his bacon, but, well, it wasn't like he was ready to focus on his remaining acreage. At least not when it came to his cattle. Maybe he was always going to be an engineer at heart. Maybe when he tapped his heart, the feeling he had for the land was a little different from what his youngest son felt. And give the father credit—he was smart enough to recognize that, smart enough to know what do when he and his ranch foreman parted ways in 1968.

The youngest son was twenty years old and still attending the University of Wyoming when his father basically gave him the keys to the ranch. As much as told him, "You're the boss; you make the decisions. Here's the checkbook—you've got five years to turn this ranch around." And then backed off.

So Jay stepped right in, commuting home on weekends during his senior year. He changed the breeding programs on the ranch, figuring out the best way to raise his own bulls. He changed the way hay was harvested. He changed a lot of things because he *knew* what to do.

But he didn't just tap his heart. He tapped his head also. He was one of the first ranchers in the county—hell, probably the state—to anticipate that computers were going to be a rancher's best friend and jumped onboard that bandwagon way before just about everybody else. You might even say he was the one driving it. He tried new things because he was young and confident and because his father stayed true to his word not to interfere.

Oh, it's not like the old man disappeared; he was plenty involved in the decision to buy more land above Steamboat Lake and expand the Fetcher's Hahn's Peak Ranch

Park Range behind the ranch

Elk Mountain to the south

Sunrise at Steamboat Lake

Moonrise over Steamboat Lake

PARKVIEW MOUNTAIN RANCH

- *Jackson County*
- *Owners: DeLine family*
- *First ranched in the 1880s*
- *Conservation easement held by the Legacy Land Trust*

Parkview Mountain Ranch lies north of the Continental Divide in North Park, near the tiny town of Rand. North Park is one of four valleys in Colorado surrounded on all sides by mountains. The other three are South Park, Middle Park, and the San Luis Valley. From the ranch, the Medicine Bow Mountains can be seen to the east, and the Park Range to the west. Some of the state's most beautiful hay meadows grow in North Park, including one owned by the DeLine family along the Illinois River. A bachelor named Laps Ish first ranched the tract in the 1880s. Rumor has it that for convenience he nailed his dinner plate to his table and wiped it clean after each meal. Many of the existing ranch buildings were recovered from the 19th-century mining town of Teller City. The DeLine family has owned and operated the ranch since 1979.

DeLine family chapel

Historic beaver-slide hay stacker

Park Range reflection

R & R LAND AND LIVESTOCK

- *Routt County*
- *Owners: Jim and Dean Rossi*
- *First ranched in the 1880s*
- *Conservation easement held by the Yampa Valley Land Trust*

For the most part, the Yampa River originates in the Flat Tops Wilderness south of the town of Steamboat Springs. It first takes on its name near the town of Yampa along State Highway 131, where it meanders through this ranch owned by the Rossi brothers. Like that of Elk River to the north, much of the valley of the upper Yampa remains in ranching. Hay meadows, views of the Flat Tops, and volcanic necks are the defining features of the valley and the ranch. Laughlin Buttes stand several hundred feet high above the Rossi hay meadows and are the weathered remains of the conduits of an ancient volcano. Marino Rossi, an Austrian immigrant, acquired property in 1923 that is part of today's ranch.

Laughlin Buttes

R & R Land and Livestock along the Yampa River

Lichens on the volcanic Laughlin Buttes

Laughlin Buttes, volcanic necks

PINEY PEAK RANCH

- *Eagle and Grand Counties*
- *Owners: Robert D. Lindner family*
- *First ranched in 1880*
- *Conservation easement held by Colorado Cattlemen's Agricultural Land Trust*

Piney Peak Ranch lies in the Sheephorn Creek drainage and is named after Piney Peak and Piney Ridge, which together form its western boundary. The creek runs through the middle of the ranch and quickly meets the Colorado River southwest of the town of Kremmling. The valley of Sheephorn Creek consists of rolling hay meadows nestled here and there in the foothills and lower ridges of the Gore Range. Elk and deer abound in this place that doesn't see much human traffic until hunting season. Aspen trees and oak brush on the ridges make a magnificent display of color in the fall. The ranch was first worked in 1880 by Canadians Joseph and Martha McPhee and their four children. At first, the McPhees settled along the Colorado River, but they moved up to Sheephorn Creek when the river valley became too crowded with other families.

Elk in winter

Ranch headquarters

Sheephorn Creek

Autumn aspens

YUST RANCH

- *Grand County*
- *Owners: Jim Yust and family*
- *First ranched by the Yust family in 1884*

Yust Ranch, also known as MY Ranch, lies in bucolic hay meadows at the confluence of the Colorado and Blue Rivers just west of the town of Kremmling. Unobstructed views exist of the Williams Fork Mountains to the east and the Gore Range to the south. However, it's the two rivers that define the ranch. Stately cottonwood trees line the lower Blue—in spring they take on the color of a lime and in autumn they become brilliant yellow and orange. The entrance to precipitous Gore Canyon is the backdrop to the ranch's west side—hay meadows give way to its steep cliffs and the sound of whitewater. There's high ground, too, that affords views of the depths of the canyon as well as the entire ranch.

Beginning in 1884, six different Yust ancestors, including Charley and his father Captain Charles Yust, "proved up" on homesteads that are or were part of the current ranch. Several ranch buildings have been preserved from the 19th and early 20th centuries. Today, Captain Charles' great-grandson Jim Yust runs the ranch.

Century-old Bill Yust cabin

Entrance to Gore Canyon

Branding irons

Wolford Mountain to the north

Colorado River sunrise

Blue River autumn

COLD MOUNTAIN RANCH

- *Pitkin and Garfield Counties*
- *Owners: Bill Fales and Marj Perry*
- *First ranched in 1881*
- *Conservation easement held by Colorado Cattlemen's Agricultural Land Trust and Pitkin County*

Cold Mountain Ranch lies along and above the banks of the Crystal River, just five miles south of the town of Carbondale. Its protected status is a reason why hay and alfalfa meadows still decorate the shores of the Crystal River and State Highway 133, despite intense development pressure in this valley not far from the boom town of Aspen. The ranch also holds an interest in the high mesa of Jerome Park, not far to the west, where hay is also cut, and where their cattle have summer access to White River National Forest-leased lands along North Thompson Creek. Marj Perry's grandfather D. R. C. Brown, an Aspen merchant who struck it big in the silver mines, began buying land in the Crystal River Valley, including what is today part of Cold Mountain Ranch, back in the early part of the 20th century.

Horsetail barley below Mount Sopris

Her nose is bleeding, her right foot hurts like hell, her kid is bellowing in terror. And, as if that isn't enough, there's this lanky guy from New Jersey wearing a battered hat with a feather who's wrapping his legs around her neck, pinning her head to a fence and trying to jam some strange pills down her throat with a giant metal syringe. Well, she might be eighty-one, but she'd show this crazy person a thing or two about how females in the West respond to *that*.

"Ohmigod, Bill!"

With a violent heave of her thick white neck, she lifts the man a foot off the ground, shakes him this way and that, and slams him into a matrix of thick wood posts, punishing his back and shoulders.

"Ohmigod, Bill!"

But no matter how hard she struggles, no matter how hard she lifts him and shakes him, the fool refuses to let go. It's like he doesn't even realize he's being thrown around; just keeps trying to pry open her mouth and jam those pills down her throat until, finally, he does.

And with that, Bill Fales uncoils his right leg from the around the neck of No. 81, releases his grip on her bloody left nostril, and leaps clear of the 1,300-pound cow as nimbly as a fifty-five-year-old cowboy can. That done, he proceeds to step into a large pile of manure, tell Olivia Pevec not to worry because he's all right, yell at his dog Malaya to stop running around and barking, and explain to a visitor, "Sulfa is a pretty broad spectrum antibiotic, but it's really good on foot rot. Usually, with foot rot you see those toes spread and I can't say that I did, but you don't always, so I'm somewhat making a shot-in-the dark diagnosis. If that's not it, then it's more of an injury in her joint, and that's when we treat it with tincture of time."

At first, he doesn't smile as he mentions tincture of time; just lets the words roll by in his usual droll way, leaving his visitor to figure it out. Sort of the way he had to figure things out for himself when he arrived out west thirty-five years ago, a long-haired kid from back east who looked like a hippie but worked like a horse. A Harvard man, no less; a guy whose father lawyered on Wall Street, whose family summered in Maine. But a guy who, it quickly turned out, had ranching in his blood just like it was part of his hemoglobin.

He starts walking toward the next chore, which looks like it's going to be haying. He can't tell for sure because high above the regal Maroon Bells and the endless sentinels of pine and spruce and aspen that make the Maroon Bells-Snowmass Wilderness jaw-dropping spectacular, high above all this beauty in an impossibly blue sky, clouds keep bumping into each other and the chance of rain keeps shuttling back and forth between likely and maybe.

"Somebody once told me, 'Only fools and newcomers try and predict the weather,'" says Fales, moving like a man who, what the hell, is going to do just that because if he doesn't get his butt over to those hayfields up on the mesa that is thirty miles away and start cutting, nobody else will. The hip-high grass whispers against his dirty jeans as Malaya, an Australian shepherd, swirls like a shadow underfoot, tracing wild parabolas of motion. But Fales doesn't seem to notice the dog. He's too busy talking to Pevec, the summer caretaker, about what needs to be done to fix her hot water heater. They walk to her cabin, where he checks out the balky appliance, writes down the needed part's number on his hand, looks at the visitor and says—deadpan, of course—"My palm pilot."

His gaze travels back to the corral where No. 81 will be sequestered with her calf for a few days, and then beyond to lush pastures and startling vistas. Suddenly, he's reminded of when "We used to get hassled by the Hare Krishnas in the airport. They'd say I was going to keep being reincarnated as a cow for every hair on the body of every cow that I butchered. I told them, I thought that living as a cow up here beat what they were doing, which was hassling people in the airport."

No doubt about it now—there's definitely a smile creasing his face as he gets into a dust- and dirt-cloaked Subaru station wagon and begins driving back down the washboard mountain road to Carbondale, to the Crystal River Valley and his Cold Mountain Ranch, to another task. Because perpetually heading to a new chore, the next urgent responsibility, is what ranchers do. Doesn't matter where they come from, it matters how much they can get done, how much they can *produce*. At least that's how Bill Fales looks at things through eyes the color of blue topaz, eyes so light they're almost transparent.

Right now those eyes are focused on the road. But his mind isn't. Instead, it's mulling over the question he's been asked about Malaya. Is her cattle-herding ability inborn or did she have to be trained?

"It's kind of like sculptors cutting an eagle or something out of a big block of marble," he answers in the kind of earnest, homespun voice that is sometimes reminiscent of the actor Jimmy Stewart. "Y'know, the eagle was there the whole time—you just gotta cut away the excess marble to reach it. You couldn't train her if she didn't have that instinct and want to work."

He keeps staring at the road, talking about this and that, driving toward more work. And, funny thing, it doesn't seem to occur to him that there used to be a time when he was the one inside that block of marble waiting to be cut free.

Bill Fales is slogging and sloshing through some of his 700 acres in high rubber boots, "moving water," shifting forty-pound boards and even heavier stones onto curtains of thick plastic tarp to alter the ditch water's direction and make the flood-irrigation process more effective. On top of his head is his signature hat, a jaunty affair with a feather that looks like a cross between something a cowboy and Robin Hood would wear, a hat that some people call "disreputable just because it's beat-up, grubby, and worn out," but which he calls "comfortable." Anyway, it beats wearing a baseball cap because "I don't see any need to use my head to advertise somebody's else's product."

Over his shoulder, rising 12,953 feet into the clouds and dominating the Crystal River Valley, is the formidable presence of Mount Sopris. He turns and walks in its direction, and soon half of his body vanishes in a sea of oat barley. A breeze unfurls, and in the late-afternoon sun the riffling grey-green plants call to mind something from the brush of Monet, one more stanza of visual poetry that isn't lost on Fales. Not only did he and his wife, Marj Perry, take the name for their ranch from a book called *Cold Mountain Poems*, both of them hold tight to their sense of wonder about their land.

"I think there's a lot of spiritual renewal just being out where we are," Marj will say later, as she prepares supper. "You just feel *good* a lot of the time."

In between popping snap peas from their garden into his mouth, her husband will add, "You can be out here at dawn irrigating, or sunset, or with the headlights when you get started late, and it's really beautiful. Or up in the Maroon Bells wilderness in October, riding on a good horse, gathering cattle off that country, and that's pretty damn spectacular."

Does it ever get old?

"Only when I'm really tired, which can happen in March and April and May and June—we probably work 400 hours each of those months," says Bill. "But when that happens, you gotta kind of slap yourself and say, 'Don't lose the wonder.'"

Of course, the older you get the easier it is to lose track of the wonder. The easier it is for your body to feel those months-that-feel-like-years and start to fall behind. As the herd kept getting bigger, and the ranch kept growing larger, "We were getting a little older, a little slower." Bill noticed things were starting to slide here and there. Mending fences. Spraying weeds. Other tasks. Definitely not good. Finally, this year he and Marj took on a hired hand, a twenty-something college graduate from Montana. He gets a house to live in for free and a decent wage. Not as much as he could get working down in Rifle for one of those energy firms, but that's okay with him. The kid likes the ranching life.

But if the hired hand helps a ton, he's no panacea for every challenge. And, lord knows, ranching's never short on challenges. Eventually, those challenges can hone the edge on a man.

Take cattle. Yes, watching calves being born is always a miracle. Yes, calves sure are cute. Yes, Fales and Perry will bring some of the newborns into their laundry room to keep them warm during freezing days in the spring. But, he insists, "They're not pets." He also says flatly, "If a cow isn't gonna calve by the first of May, it doesn't have a home on this ranch. Some people say, 'Oh, but she's a great cow, you *gotta* keep her.' But my definition of a great cow is one that produces me a live calf every year. If a cow doesn't calve, I don't want that genetics in my herd."

The jaw that looks like it was chiseled from Mount Sopris is sticking out, not quite defiant but sure as hell unrepentant because out here there's not a whole lot of room for sentimentality.

"I think ranchers have a truer, or more honest, relationship with the outdoors. We see the beauty, but we also see the harsh reality—we see death much more than people who don't work with animals do. Maybe people working in a hospital see death, it's a constant with them, too. You see it from disease, from coyotes, from an injury, from lightning, from suffocation when they're born. I mean, things can die all the time, and because we work with animals and stuff, we see a lot of death. If you're living in town and working at the lumberyard or somewhere, death is a very remote concept for America these days. But not for us."

A few years back, in fact, Fales became so intimate with nature's harsh reality, it felt like a two-by-four smacking him in the face. "Open cows" is what they call it, cows that can't breed. Cows that eat your grass but give you nothing in return. Cows that don't deliver on their end of the bargain. Fales looked for answers everywhere. Was it some kind of venereal disease? Was it a mineral deficiency? He never found out for sure. All he knew was he had to go out and replace thirty-five percent of his herd, selling off the open cows, acquiring fertile ones in a hurry. So much of a hurry that "It was a case of classic ranching—buying high and selling low."

For a minute, there's a sort-of smile on his face, but even that's about fifty percent rueful.

That smile hints at other realities to be confronted in the Crystal River Valley. Not as harsh as death or herd sterility, but just as final and pretty damn painful to a rancher's soul. Realities like the tide of development that has washed over the land, land where "a lot of sweat, blood, and broken bones" had been mulched into the soil only to be covered up, laminated with an eighteen-hole golf course and something called the River Valley Ranch, which isn't a ranch at all, unless your idea of a ranch is a mosaic of million-dollar trophy homes.

Not all of the valley's ranchland has been turned into manicured parcels; a lot of it has stayed intact, even if most of the family ranches that used to dominate the landscape are gone, folded up into giant operations. Whenever the land changes owners and the familiar calloused hands that worked it go away, that causes a sadness in the community of ranchers. And for those who remain, there is still the glum reality of seeing land where they grew up—or came of age—sold for an ungodly amount of money. Because in the Crystal River Valley, and the nearby Roaring Fork Valley all the way from Aspen to Glenwood Springs, real estate values keep exploding so fast and so far that $30,000 for one acre isn't crazy, it's right in the ballpark.

But Bill Fales doesn't want to play in that ballpark. Doesn't like the game or the rules it's played by. Big money doesn't necessarily yield good emotional dividends. Even if you don't ask him, he'll tell you that. He'll tell you, "I've seen throughout my life lots of people with too much money, and I've yet to notice that they're any happier than anybody else."

For him and his wife, money won't bring back the way the land looked or felt when Marj's grandfather D. R. C. Brown, an Aspen merchant who struck it big in the silver mines, began buying it up back in the early part of the 20th century. Actually, in those days, there was more farming than ranching going on; the valley was a hub of potato farms.

Marj's grandfather even paid for his land—some 1,400 acres—with the money from one year's crop of spuds.

By 1941, potatoes were still the big item, but that was about to change. Ruth "Ditty" Brown, D.R.C.'s daughter, had married Bob Perry and they moved into an impressive, two-story Victorian home that rose up on the property. Bob grew potatoes for a few years, but then he focused on ranching. The Mt. Sopris Hereford Ranch began to expand: Bob could buy land for as low as twenty dollars an acre, and on that land he built a herd of cattle that grew to as many as 400.

As Bob and Ditty's ranch increased, so did their family. By the time they were done, uh, procreating, they had seven kids, and all of them grew up ranching. But only two were boys, and there were seasons during the year—summers especially, when there was haying to be done—that extra pairs of arms didn't hurt. So Bob got into the habit of hiring a few young men. Mostly, it worked out fine. The men worked hard, got a place to stay and money in their pocket, and Bob Perry got the help he expected. But in 1973, he wound up getting more than he expected when he signed on some kid from New Jersey with shoulder-length hair. Bob didn't know it then, but he'd just hired his future son-in-law.

Darlin' Willie—as his mother called him—was the youngest of Hal and Katherine Fales's five children. He was born in Morristown, New Jersey, but grew up in rural Gladstone in an old farmhouse owned by his grandparents, surrounded by 500 acres. His father may have taken the train to Wall Street every weekday morning, but Bill grew up loving country life, an energetic kid who played a lot of hockey, did well in school, and only got into trouble when he'd mope through the house, grumbling to his mother, "There's nothing to do around here." Chances are he got some of his energy and love of the outdoors from his grandmother, who was into farming in a big way. Dorothy "Dodo" Fales was also a damn good horsewoman, handling herself with ease on jumpers or during fox hunts. And although she was married to a New York City banker, she wasn't afraid of hard work. After all, hadn't she lugged some of the rock used for the half-mile driveway that led up the farmhouse all by herself?

The Fales were comfortable but frugal. Katherine considered Coca-Cola a luxury—much to her youngest son's dismay—and her suppers often featured hearty but inexpensive fare like corned-beef hash and Welsh rarebit, which the kids knew was just a fancy way of saying melted cheese on bread. But Hal did well enough to buy half a share of a house in Maine, where the family spent summers. He also managed to send all his kids to good boarding schools; in Bill's case, it was Middlesex School, in Concord. Between Massachusetts and Maine, he grew up with a "strong dose of American Revolution ideology."

In 1971, Bill began attending Harvard, but living so close to a big city like Boston didn't feel right. He took a year off and worked in Maine as a carpenter. He had always liked Maine, in part because of Ed Smith, a half-Penobscot Indian handyman who used to let Bill tag along while he made his rounds. Thanks to Smith, young Bill learned to shingle roofs, fix stoves, and solve plumbing and electrical problems. There was nothing too tough for Smith to tackle. It was like he'd tell his young charge, "Some fellow was smart enough to put this together; we ought to be smart enough to take it apart." Then he'd do just that, fix the problem and put whatever he'd taken apart back together again. Bill watched and learned and eventually came away with both a skill set and an attitude. Years later, as a rancher, he wound up fixing just about everything he needed to. He'd even say that every vehicle he owned—truck, tractor, baler, you name it—never had to be sent out to someone else's shop to be repaired. Built his own house as well, for that matter.

A year off from Harvard didn't make Bill miss it. It was 1973. He was twenty years old, antsy, and coming down with a case of wanderlust. Through a friend of his family, he got the name of a rancher in Colorado who was hiring hands for the summer. So he packed up his used pickup truck and off he went—and with his parents' blessing, too. Katherine Fales had always told her son that, as far as she was concerned, he could be anything he wanted to be as long as it wasn't a businessman.

He didn't know it at the time, but waiting for him was a spirited girl who was about to change his life. Her name was Marj Perry, and she was one of those seven children belonging to Bob and Ditty. Like all of her siblings, she was nurtured on ranching, learning to ride early on, becoming no slouch with a branding iron. Every year, in late June, she'd help drive her father's herd out their front pasture onto Highway 133, then along Highway 82 to Snowmass Creek and up to the summer grazing area that was known as Kate's Lot, named for the Danish lady who had homesteaded it.

The thirty-mile trip took three easy days, and the only headaches were the ones created by bovines. The Crystal River Valley was pretty sleepy back then; there was none of the rush-hour traffic that would afflict it years later, when tourists and commuters began jamming up Highway 133. Carbondale itself only had about 700 people and more than a few unpaved streets. Not that asphalt was in great demand. When Marj took her driver's license test, she had a hard time finding any cars on the street so she could parallel park.

Carbondale and the valley got a whole lot more interesting for her when the long-haired guy with the intense blue eyes drove up to the ranch. He was pretty cute, but pretty raw when it came to high-country agriculture. Still, he was willing and hard working and a lot about ranching came quickly and naturally to him. But not everything. Take the first time he watched the vet perform a Caesarean section on a cow. Years later, Marj would insist that Bill fainted. He'd deny that, but concede, "I turned green."

But he kept at it. A "less than passable" rider, he learned to become an accomplished one. He learned about haying, about how to cut it, how to read its moisture content, how to bale it. He learned how to fix fences, how to clean irrigation ditches, how to change water. Eventually, he learned about cattle, how to take care of them, how to help pull calves out of their mommas when the natural birth process went a little haywire. He also learned about falling in love.

The courtship was natural, although not necessarily gradual. When Bill went back east that fall to give Harvard another shot, Marj went with him. After that didn't work, they came back to Colorado together. Both attended the University of Colorado in Boulder and earned degrees, he in political science, she in geography. Eventually, they got married, even though Marj kept her maiden name. Sure, that was a little curious for the valley, but, well, these were the seventies; Marj wanted to make a "statement."

They had kids—first Molly, then Katie. They became more entrenched in the valley and its goings-on. Bill became a fixture on Pitkin County Open Space and Trails Board, serving for thirteen years, five as commissioner, working hard for open-space conservation programs, showing up at meetings in his "trademark cowboy hat, weathered, sweaty, and anointed in the grit and grime of a way of life that he so passionately seeks to preserve." At least that's what it said in the county proclamation that honored him. He even entertained the notion of running for county commissioner—lord knows he'd been outspoken at enough meetings when he felt ranchers were being squeezed by government and development—but thought better of it. He had enough on his plate.

He continued working for his father-in-law, working his way up to a whole $600-a-month after nearly twenty years. To be fair, Marj and Bill did get a free home on Perry land, a free milk cow, and free meat—as long as they butchered it themselves. But early on, although their informal partnership with Mt. Sopris Hereford Ranch would endure for decades, they had begun to move toward independence. In 1979, they bought 3 heifers and began slowly expanding their own herd of Hereford-Red Angus crosses (with some Gelbvieh blood thrown in), which would grow to a little more than 200 over the decades. In 1983, they obtained a lease option on a 160-acre parcel that abutted Bob Perry's land, and was owned by Mike and Pauline Desandre, longtime ranchers in the valley. Two years later, they bought it. In 1990, they purchased 420 acres from Bob Perry himself. Over the next few years, they added another 100 or so of his acres.

But all along it wasn't just about having their own land and livestock. It was also about having those spectacular views up near Kate's Lot, especially when they brought the cattle down in the fall and the colors were like fireworks. It was about having the constant companionship of Mount Sopris and the unshakeable feeling that it was their own private mountain. It was about knowing that most of the people they chose to deal with valued honesty and a handshake-deal as much as they did. It was about the freedom of being outside all day, working at their pace, nobody telling them what to do or when to do it. It was about the freedom of independence, embracing a grueling work ethic the way they wanted to, figuring out for themselves that "you can't raise cattle by shooting the bull."

Nope, it wasn't Utopia. Over the years, they had to deal with some of the environmental extremists, people who, according to Marj, "Think ranchers are just a bunch of Neanderthals who ride around on horses and ruin the environment." For Bill, it was just plain "aggravating" to have "these people, these rabble-rousers, tell you you're raping the land. That the West used to be this idyllic place with wild animals, wolves and grizzly bears, and lots of grass and no erosion. No environmental problems whatsoever until ranchers came along." What the hell were these people thinking? Any rancher worth his salt lick knew that his relationship to the land was a "partnership. You have to take care of it, so it takes care of you. If you abuse it, it'll quit giving you any return."

The Colorado girl and the New Jersey boy also didn't appreciate being squeezed by government regulations, challenged as to what they could or couldn't do. Sometimes it all got to be so much it just detonated Bill, made him just want to say, "This is my land. I own it. Get the hell off it!"

Ranching was tough enough as it was. Sure, you had the freedom, but you also had to confront the seesaw feelings of control and lack of control, the uncertainty of what soaring corn and fuel prices would do to your bottom line, what an epidemic of open cows would do to your herd. But mostly they figured they could figure anything out. Somebody, or something, had put the problem together. Surely they were smart enough to take the problem apart, fix it, and put things back together the right way.

And at the end of another day, when shadow-wreathed Mount Sopris loomed as a guardian phantom and those nearby sandstone cliffs were like blood-red tidal waves frozen in the twilight, when the water was coursing through the ditches and the cattle looked healthy-plump and the hay was growing, well, that was a pretty good payoff. True, sometimes the payoff was about the money because money was a "good tool; it helps you do a lot of things. You have to be on economically sound enough footing so you can take a long-term view and not abuse your land for a short-term gain."

But it was never just about profits. No, there was always something else they got from the ranching life, something Bill put into words easily and effortlessly when he said, "This fits my sense of what's right. I'd have a hard time being an advertising agency guy or a developer. I like taking care of the land. I like working with cattle, I like producing food. There's a real sense of accomplishment in that. You want to know why I ranch? Because I *love* it."

They loved their work, they loved their children, they loved their land. Yeah, in some ways it seemed that Bill and Marj had everything they could ever want.

They just didn't get to keep it all.

Over the decades, Bill Fales had felt different pulls and changes in ranching. Mechanization had come—larger tractors, large hay balers—making it easier for one man to do the work of several, even though that one man had to work harder than ever. Improved knowledge of genetics and breeding made the herd more efficient, but it also made a rancher's responsibilities more complex. And, speaking of complex, why sometimes it felt like you had to be equal parts MacGyver and veterinarian to succeed as a rancher. Mindful of this, Bill was fond of quoting prairie poet Baxter Black's definition of a cowboy: "Someone who can replace a uterine prolapse in a range cow in a three section pasture with nothing but a horse and a rope."

But maybe the biggest change of all was that agriculture seemed to be disappearing from the area. The growth of the valley—a growth Fales labeled "turbocharged, exponential, logarithmic, whatever you want to call it"—had been accelerating for decades, consuming ranchland and farms, belching out commercial and residential enclaves. In 1945, Pitkin County—the county in which Cold Mountain Ranch would one day sit—had 93,135 acres of agricultural land. Sixty years later, that figure had dwindled to 44,450 acres, less than half. As the acreage disappeared, so did the people who worked it. By the end of the 20th century, what had once been maybe fifty self-sustained operations had shrunk to what Bob Perry called "practically zilch." His son-in-law, trying to be a little more precise, reckoned there were only four families in the county who depended on ranching for their income—"the Jeffersonian concept," he called it. Most of what agriculture was left consisted of sprawling operations. Sure, these kept livestock and hayfields on the land, but, well, somehow they weren't cut from the same fabric as the old ranching community. But far worse than the slow evolution of Big Ranching was the onrush of development that began to overwhelm rural life along Highway 133.

An uneasiness that had been growing inside Bill took icy hold of him in the early 1990s, when good ag land just outside Carbondale was gobbled up for what would become the River Valley Ranch. Even Bob Perry had sold 120 acres of his property to developers. Made good money, too. Hell, developers were willing to pay ridiculous prices for land because the people who were flocking to the valley were willing to pay even more ridiculous prices. By 2008, a *town house* around Carbondale was being shopped for over $800,000.

But money had never held that much appeal for Bill Fales. He looked around the valley and wanted its beauty, its "tranquility," preserved. He wanted the land to retain its utility, its productivity. He looked around and thought about California, where untold acres of fertile farmland had been sacrificed at the altar of development. He knew he couldn't let that happen here. At least not without a fight.

The first conservation easement he worked on, finalized on the last day of 1996 and recorded the following spring, was a complex and collaborative effort between seven families—including his—that owned a 4,800-acre wedge of land. But that wasn't the end of Bill's involvement in the easement process; it was just the beginning. The threat of development wasn't going to go away; the land had to be protected, saved from being homogenized into trophy homes and golf courses. And not just for Bill and Marj's generation of diehard ranchers. On more than one occasion, they would cite the old Native American proverb that went, "We don't inherit the land from our ancestors, we borrow it from our children." They worried about what would be left for Molly and Katie. The girls had always gotten a "healthy dose of propaganda" from their parents about ranching, but Bill and Marj didn't know if their daughters would ever want to work the land. What they did know is Molly and Katie deserved the opportunity to be able to make that choice when they were ready to.

No, easements weren't foolproof. Bill knew they couldn't "mandate productivity. I don't know of any way to mandate that a guy gets out of bed at five-thirty in the morning to change water or feed his animals. All they can mandate is the potential that the land can be productive." But they were a damn sight better than standing around watching your legacy shrink to nothing.

Slowly, carefully, sometimes painstakingly, they obtained other easements. In 2006, they protected another eighty acres they owned. Then they began what would be a slow, at times torturous, two-year process to place all of their remaining acreage under easements. They also worked with others in the valley to protect land they didn't own. Along the way, Bill became president of the Colorado Cattlemen's Agricultural Land Trust and worked to spread the gospel of easements to other parts of the state, to help maintain the potential for tranquility and productivity.

No, he didn't feel like he was on some noble crusade—there was a quid pro quo involved, because conservation easements often came with cash subsidies from conservation groups or local government. But these subsidies—while very much welcomed and, in some cases, very much needed—usually didn't amount to stratospheric numbers. At least not when you compared them to the sums being offered to some ranch owners by private buyers. And sometimes, no matter how much they loved their land, those ranching families just couldn't walk away from so much cash. Which was a lesson that Bill and Marj learned personally and painfully.

Over the years, they had acquired those 520 acres of Bob Perry's ranch. On that property, they designed and built their home, raised their two daughters, expanded their cattle operations. For Marj, that land, and all the rest of her family's acreage, was part of her heritage. It was where her mother and father had raised her and her six siblings, where she had come of age, where she had met her future husband. For Bill, it was where he had been baptized into ranching, where he had turned green at his first bovine C-section, learned how to hay and be a more than passable horseback rider, learned to figure out a lot about who he was and what he was made of. Both of them wanted the land, and the sense of wonder it could evoke, to stay in the family.

But the family felt otherwise. Or, as Bill would recall, "Basically it was me and Marj on one side and everyone else on the other side. We didn't have much public support for our position."

By 2006, Bob Perry was eighty-eight and in failing health. He and Ruth and six of their seven children lined up along the idea that selling the 1,180-acre Mt. Sopris Hereford Ranch was the best plan for the future. And on August 18, that's what the Perrys did, signing a contract to sell their property for $27.25 million to Sue Rodgers and Tom Bailey, two very wealthy landowners in the area who ran significant cattle and horse operations, respectively. Rodgers acquired more than 800 acres, Bailey picked up about 300. The remaining 29 acres went to Bill and Marj. Two days later, Bob Perry died at his ranch, leaving behind a wife, seven children, twenty-six grandchildren, and twenty-nine great-grandchildren.

Bill and Marj had purchased those 29 acres, which adjoined their Cold Mountain Ranch, for $644,000 from the Perry Family partnership. They had wanted to buy more. In fact, they thought they had an understanding with Bailey for an additional 154 acres from his parcel, but that turned out not to be the case. They also thought that Bailey was going to preserve the century-old two-story Victorian house where Marj had grown up, as well as the historic barn, both of which now sat on his purchase. But that also turned out not to be the case. In July 2008, two years after he laid claim to his piece of the Perry land, Bailey had both buildings razed, and a little piece of Marj and Bill died.

Steely indignation and disappointment percolate in Bill Fales's voice as he drives by the jumble of weathered wood posts and beams that were once Bob Perry's barn, an iconic building, a barn the *Aspen Times* had called "regal."

"That just makes me sick," he says, almost spitting out the words. "I didn't give a damn about the house, but the barn? It had timbers that were sixteen-by-twenty-four! Great stuff! It could have been saved and rebuilt elsewhere, but they just reached in there with a backhoe and bashed it down."

Bill isn't kidding about salvaging and saving the barn. Less than a year ago, when Bailey was going to get rid of an old log cabin on his property, he gave Bill three days to figure a way to lay claim to it. No problem. "I jacked it up, got some big steel I-beams and slid them under it, and pulled that onto a trailer and brought it home."

There is a delicious pleasure in his voice as he talks about saving the cabin. But even that memory isn't enough to melt the cold stare in those pale eyes, or stop him from slowly shaking his head as he stares at another piece of history that has been shattered into splinters.

The good news is that he doesn't stay pissed off for long. There's too much beauty around him, too much life, too much wonder. "Look at that little colt there," he says, his enthusiasm reborn, his voice infused with a golly-gee-whiz tone a la Jimmy Stewart. He looks out the car window at the animal, smiling. "Probably just a few days old."

The station wagon keeps moving. He's heading toward the Village Smithy restaurant in Carbondale. Wants his visitor to try one of its hamburgers, which are made with his ranch's beef. "Best hamburger you'll have eaten in a long time," he says matter of factly—and correctly.

When lunch is over, he wanders out into a now-sunny afternoon. The army of dark clouds that had bivouacked in the sky and seemed ready to drench the land have retreated and Bill's been blessed with the freedom to handle another task. This one takes him up to a pasture belonging to one of Marj's brothers. In exchange for taking care of his brother-in-law's haying, Bill gets to keep half of the crop. He climbs into the enormous New Holland HW300 tractor he bought used in Wyoming ("It was a steal") and lumbers off at 3.7 mph, pulling the swather with the sickle-bar mower behind him, patiently navigating his way up and down the field as the combination of alfalfa, orchard grass, and brome is swiftly cut and left in neat five-foot-wide rows.

The wind is making the grass sway, cloud-filtered sunlight comes tumbling down, and the whole scene is looking like another Monet, but this time Bill Fales doesn't seem to notice. He's got other things on his mind. Maybe he's hoping that, when it comes time to stick a probe into the rows of grass, it will indicate a moisture content of eighteen percent, which means he'll be able to bale "perfect hay." Or maybe he's thinking about the irrigation water he'll be moving down on his ranch in a few hours. Or maybe he's wondering about the damaged foot of No. 81 and whether or not the sulfa and the tincture of time will heal it.

Yeah, he's got a lot on his mind, but that's okay because he gets to think about it outside. At his own pace. Without anybody squawking at him or telling him what to do or when to do it. He'll just keep on working, producing, doing what feels right. Being outside.

He steps down off the tractor, picks up a piece of fresh-mown hay and puts it in his mouth, slowly chewing it, almost like *he's* grazing on it. He leans against the tractor, arms folded, and looks around with eyes that are now the color of a late-afternoon sky. That disreputable hat with the feather is pushed back on his head, revealing silver hair that's sweaty and matted down. He doesn't say anything, but then, he doesn't have to. He's too busy feeling free, feeling right, feeling productive.

Then he climbs back in the tractor, kicks it into gear and gets back to work, looking so at ease and doing it so naturally that it's hard to imagine there was ever a time when it was all new to him. It's hard to imagine what it was like thirty-five years ago, when a long-haired kid from back east showed up in the Crystal River Valley looking for a summer job, not realizing he had just found a calling. Not having a damn clue that he was really a rancher trapped inside a block of marble, just waiting to be cut free.

Hay meadow below Mount Sopris

LOST MARBLES RANCH

- *Pitkin County*
- *Owners: John, Laurie, Johno, Peter, and Kate McBride*
- *First ranched in 1880*

Lost Marbles Ranch is tucked up against the White River National Forest and below the snowcapped peaks of the Maroon Bells-Snowmass Wilderness, not far from the city of Aspen. Visible to the west is 12,953-foot Mount Sopris; 14,130-foot Capitol Peak can be seen to the south. Lost Marbles Ranch is one of only a few remaining working cattle ranches in the Aspen area, most having been lost to the pressure of development around the recreational mecca of Aspen. The valley of East Sopris Creek was originally homesteaded in small parcels by numerous families, including Kinney, Staats, and Light, as early as 1880. The Light family raised saddle and Belgian horses. Johno McBride and his family live in a house built in 1881 by homesteader Daniel Fealy. There was an attempt to develop the ranch for homes in the 1970s. The McBride family began assembling the ranch in 1979 and restored it to the condition of a working ranch. East Sopris Creek courses the ranch and provides water to feed its gravity sprinkler system used to irrigate hay meadows.

Kate McBride's horse ranch below Mount Sopris

Kate McBride's barn

Johno McBride's barn below the Maroon Bells-Snowmass Wilderness

Gravity sprinkler system

HARVEY RANCH

- *Pitkin County*
- *Owners: Connie Harvey family*
- *First ranched in the 1890s*
- *Conservation easement held by Pitkin County and Aspen Valley Land Trust*

Like it's neighbor to the north, Lost Marbles Ranch, Harvey Ranch is tucked up against the White River National Forest and below the snowcapped peaks of the Maroon Bells-Snowmass Wilderness. At 12,953 feet, Mount Sopris is conspicuous to the northwest from the ranch's higher parts. Massive aspen forests, sage and grass meadows, and remote stock ponds serving the cows and horses that live on the ranch define its landscape. This is a very colorful area, especially when wildflowers bloom in summer and the aspens turn yellow, orange, and red in the fall. This portion of what is today Pitkin County once served as a Ute Indian reservation. The Meeker [Colorado] Massacre of 1879, during which many U.S. soldiers were killed by Ute Indians, caused the closing of the reservation, and the land was subsequently homesteaded. The Harvey family acquired the property in 1962.

Mount Sopris

Sunrise on Mount Sopris

VOLK RANCH

- *Gunnison County*
- *Owners: Volk family*
- *First ranched in 1911 by the Volk family*
- *Conservation easement held by Colorado Cattlemen's Agricultural Land Trust*

Volk ranch lies on a mesa that's not really visible from anywhere, but is below one of the most unusual mountains in Colorado, Ragged Peak. Its eroded granite fluting makes it different from any other mountain in Colorado, except for Marcellina Mountain next door. Both are laccoliths—granite mountains formed when igneous domes swelled the crust of the earth—that were exposed when the surrounding rocks eroded away. Volk Ranch itself is defined by the verdant pastures covering its rolling hills.

Slovenian immigrant George Volk Sr. left the coal mines in the town of Crested Butte in 1911, crossing Kebler Pass with a milk cow, a beef cow and calf, and a wagon, to realize his dream of having a ranch. He first raised grains and vegetables to feed the miners in nearby Somerset before starting his cow herd, and he, himself, continued work as a miner in the winters to make ends meet. His original cabin stills stands, and his son George Jr. and grandson Gary operate the ranch today.

Sunset on Ragged Peak

TRAMPE RANCH

- *Gunnison County*
- *Owners: Bill and Dora Mae Trampe*
- *First ranched in the late 1870s*
- *Conservation easement held by Colorado Open Lands*

Trampe Ranch consists of several unconnected parcels, all of which lie in the valley between the towns of Gunnison and Crested Butte along some of the most magnificent rivers in Colorado, the East and Taylor, and Brush Creek. If that isn't enough scenery to stimulate one's senses, each parcel affords views of either the Elk or West Elk Mountains. Massive hay meadows, pristine wetlands, aspen forests, and some classic barns add to the sublimeness of Trampe Ranch. Henry Trampe began acquiring land in the valley in the 1890s, first growing potatoes, then raising hay and cattle. His grandson Bill took over the management of the ranch at age twenty-one when Henry's son and Bill's father, Sheldon, died of a heart attack. The Trampe Ranch lies in the midst of some of Colorado's most beautiful public lands, and chances are good that a visit to the Crested Butte area will yield a sighting of Bill Trampe tending to his cows, water, and meadows.

Gothic Mountain

Green and mucky and launching a stench so foul it could bend your knees, the cocktail of cow dung and urine is being diligently shoveled from the cattle trailer by a lanky cowboy with bad shoulders, worse knees, and a once-broken neck. Under a battered baseball cap with CAT printed across the crown is a lake of silver hair lit to incandescence by the morning sun. Under that shimmering hair is a sixty-one-year-old face that is so rawboned, tan, and rugged, it looks like it belongs on the Marlboro Man—which, in fact, its owner was *this* close to being.

But the never-ending aches in his joints and the memory of what might have been are as inconsequential as tumbleweed blowing across the prairie. No, right now, all that matters to the cowboy is the shovel and the shit and the task at hand. He knows that he better get this job done because, sure as shooting, there's another lurking nearby. And another one after that. So all you hear for a while is the scrape of metal on metal and the slow slosh of waste from the trailer. Then you hear something else.

"Hey, remember you asked me about the romance of ranching?" he says, a god-dam-big smile creasing his face, the shovel slowing down as he nods his head toward the trailer floor. "Well, here it is."

Okay, so the first thing you should know about Bill Trampe is—and this will undoubtedly shock the hell out of a host of non-native county officials, recreation zealots, and citified residents in the East River Valley—he has a sense of humor. The second thing you should know about Bill Trampe is it's a good thing he does.

Not that he makes that sense of humor so readily apparent. It's hard to see a lot of the time, especially when he's talking about being badmouthed and disrespected and feeling like a "Goddam endangered species" right in the place where he's lived his whole life.

Sure, his blue eyes have more than a sparkle or two left, but sometimes they look tired. Yours might, too, if they'd seen what his have. If they'd seen the very last cattle train pull out of Crested Butte, or the parade of hay pastures that used to shimmer nearby until the powers that be built a dam and flooded them into oblivion. If they'd seen fifteen ranches dwindle to four and the proud heritage of working the land get swallowed up by the hungry lust to "thirty-five" the meadows and graceful hills into trophy homes so pristine and Disneyland-like it's enough to make you want to spit. If they'd seen the *looks*, the icy, hurtful stares from people in a restaurant where once you were welcomed but now are shunned just because you're wearing weathered rancher clothes, and maybe you have a little bit of manure on your boots—which is what a man herding cattle all day is supposed to have on his boots.

Maybe if you'd seen all this and heard the curses and the dismissive comments you'd have to pat all your pockets before you could find the place you keep your sense of humor. Hell, maybe it wouldn't take thirty-odd years of this to sour you. Maybe all it would take was the tight, crazy grip of some man's hand around your shirt collar as he dragged you beside his fancy car and then pushed you down and ran over your leg as he sped away. And why? Because you had the gall to drive your cattle across a road you had every right to cross, only he wasn't about to stop for a bunch of filthy cows or their dumbass owner. Maybe you'd be as crusty as an apple pie too, and not nearly as sweet, if you could look a stranger in the eye and say, "Being crosswise is a way of life; I don't like it and I wish I didn't have to be there, but that's the way it is."

There's no self-pity in Trampe's voice when he says this. Or even when he says you're damn right he can envision a future West without ranchers and "Frankly, it sickens me." There's no self-pity mainly because he has no time for self-pity. Not when there's irrigating to do, fences to be mended, cattle to be worked, hay to be baled and—maybe most important of all—land to be saved. It's no wonder he's never skied, never fished, never bicycled. Just worked. Worked so much his body's shot to hell. Not that there's much body left. Not unless you call one hundred and fifty pounds stretched across a six-foot-one-inch-frame much.

"Since my early adulthood I was busy having to run this ranch and keep it afloat," he says, explaining that outside pursuits "didn't seem to be proper for me."

Then, there it is again—that easy smile. And then that loud, throaty laugh. And suddenly the gloomy tone is gone and Bill Trampe looks up at you, leaning on his shovel, and says, "Guess I'm pretty boring, huh?"

Not hardly.

Of course, when you're talking about the Trampe Family, it's not as if you're talking about a bunch of loudmouthed egomaniacs or life-of-the-party cutups. No, what you've got are salt-of-the-earth types; clay and gravel and chunks of granite mixed together in a kettle full of gumption. No-nonsense folks who never had a problem doing things the hard way if they had to, even if part of that way was straight uphill.

Henry Trampe was the oldest of eight kids and probably the sickest. As a boy growing up in Metropolis, Illinois, he couldn't do any of the farm work his parents and siblings did. After college, the doctors told him he had tuberculosis and needed a change of climate, some place drier. For Henry, that turned out to be Pueblo, Colorado. At least for a while. After teaching school in hamlets like Cotopaxi and Howard, he grew restless. So he got on his bicycle and headed west over roads that were a long way from paved, stopping here and there, reconnoitering, absorbing, breathing in the glory of what he saw. About three years later, sometime in the 1890s, he pedaled and pushed that bike over Marshall Pass and rolled smack dab into Gunnison.

His health had come back and at last he was free to do what he'd always wanted to. North of town he found himself some land that wasn't much more than sagebrush and rocks and turned it into a potato farm. Over the years, he added some cattle. Then some children. His only son, Sheldon, grew up taking to the ranching life, and slowly the Trampe property line began extending. The land was still stubborn, but over time it yielded to hard work. Eventually, the potatoes were phased out in favor of hay and more cattle.

Sheldon married a little late in life—hell, he was forty—but once they tied the knot, he and the former Dora Mae Entz set about starting a family. First Bill, then Bette Mae. Like his old man, the son just took to ranching. He loved the land, loved the way it opened up into valleys and basins south toward Powderhorn, where three of the best ranches you ever saw were located until they got swallowed up by the Blue Mesa Reservoir. He loved the land all the way north to Crested Butte, too—which is where he happened to be visiting one day in 1951. There he was, no more than four years old, holding his mother's hand when the last cattle train—and those were Trampe cattle, too—pulled out of Crested Butte, bound for the Denver stockyards, chugging into history, never coming back.

Pretty quick, Bill was helping on the ranch. Even when he only weighed no more than seventy-five pounds, he'd be out there helping. If he couldn't lift those ninety-pound bales of hay, well, at least he could roll them along the ground. The older he got, the more work he did. He became active in 4H and Future Farmers of America. Didn't have to be pushed, either—ranching was just *in* him. He was an "outside guy." A "hands-on" guy. Never gave a second thought to what he wanted to do when he grew up. Never worried about his career path.

But maybe his father did. Sheldon began to sense a change shivering through the East River Valley, and it was coming from the north. Crested Butte—not twenty-five miles away—was on life support. Just like the gold and silver mines before them, the coal mines had run their course; by 1952, most were shut down. Agriculture was all the valley had, but that was about to change.

About 1961 or so, the Malensek Family sold its ranch to a company that was going to build a ski area in Crested Butte. It didn't seem like a big deal at the time, but Sheldon saw the future quicker than most. He told his fourteen-year-old son, "You'll be lucky if you can finish your ranching career up here."

The boy had heard a lot of things from his father. A lot of advice, a lot of sayings. Stuff like, "Make sure you're tunin' in before you start broadcasting." In other words, listen before you talk. That made sense—a lot of what Sheldon said did. But all this predicting, this talk about the death of ranching in the valley? Well, Bill would just have to see about that.

So the father's grim vision of the future didn't get in the way of the son's hopeful one. Bill's life seemed like a natural progression, growing deeper and deeper into the land. Hell, there'd always be agriculture—that's what the land was here for. He'd be running the ranch one day, he knew that. Everybody knew that.

They just didn't know it would happen so soon.

It was 1967, and Bill was away at Colorado State University when the call came. Sheldon had been in a ditch, irrigating, when his heart quit. Just like that. No warning. No medical history to speak of. He was sixty-three and he was never going to get a day older.

School got pushed to the back burner—it would never return to the front—and Bill came home. He was barely twenty-one, but now he was the man of the ranch. And he was ready. He'd always been that hands-on guy, and now his hands were full. He started moving the ranch in directions he wanted, directions his father hadn't gone. Like changing his cattle, slowly producing heifers that could be successfully bred as two-year-olds. In order to augment the genetics, he had to focus on their nutrition—and that meant taking a different look at what was growing in the Trampe hay meadows. Buying feed was too expensive; he needed to be self-sufficient when it came to feeding his animals. So he improved the irrigation system, began growing more legumes, more clovers, more alfalfas—hay with a higher protein. He began to cut it earlier because he knew grass is more nutritious when you do.

He labored like hell, caring for the animals, caring for the land. Didn't even get a paycheck—his mother didn't see why he needed one when he got room and board. That would persist for the next twenty years or so. Dora was strong-willed. So was her son. If things weren't always rosy between them, they both agreed on one thing: If the ranch was going to succeed, hard work was the only way.

But he didn't mind that one bit. Matter of fact, he welcomed it. An acceptance of—maybe even a hunger for—hard work, bone-hurting work, had been bred into him; it was a part of him, an important part. So important that when he had an opportunity to become the Marlboro Man, he went and sabotaged himself. That happened sometime in the 1980s. Executives from the company were scouting the Gunnison area for possible filming locations. Word came down through friends of friends that they were also looking for a new model, some rugged cowboy to be the face of their smokes. Bill sure had that look. Strong jaw. Blue eyes. No-nonsense as rock. Was he interested?

He wasn't sure, but he agreed to an interview. Hmm. The more he thought about it, the less sure he became. Yeah, the money would be good, damn good, "outstanding" good. Good enough to take it easy, to quit having to get stomped on by cows and horses and get his damn neck broken like he once had only he didn't find that out for years. Good enough to not have to wrench his shoulders throwing around those ninety-pound bales during haying season. To not have to worry all the damn time about how to make the land profitable enough to live on.

But he also knew that if he got the job, he'd be at the beck and call of the Marlboro people. Have to go where they said when they said. He knew you couldn't run a ranch that way. He also knew "there's more to life than money." Especially when you're a hands-on guy, an outside guy, an independent guy.

Still, he'd given his word about the interview. So he showed up. Only he went wearing the "cruddiest" clothes he could find. And he rode in on the "sorriest-looking horse" he owned. And, no sir, he wasn't surprised when the Marlboro people said thanks, but no thanks. Wasn't disappointed, either.

So he went back to ranching, back to the life he was comfortable with. Only thing was that life was changing. And not in a good way.

It may have started in the 1960s because of the ski area. It might have started anyway, even without the ski area. Whatever. Over the years and the decades, in Gunnison and heading north along land that was nourished by the tumbling Taylor, East, and Gunnison Rivers, development began to take hold. The beautiful vistas along the East River Valley became a magnet for fishermen, hikers, mountain bikers; people who sure weren't ranchers, people who got in Bill Trampe's way. People who would leave open gates on federal land where he had grazing permits so the cattle would get out, some dying when they ate poison grass. People who would be aghast when they discovered they had to hike or pedal through manure—or cows themselves—and would scream, "Move the goddam cows, you goddam cowboy!" (And you're goddam right, he'd curse back.) People who didn't know those goddam cows had created those trails. People who didn't seem to respect NO TRESPASSING signs because they thought the pristine meadows around Crested Butte should belong to people, not animals, and barbed wire fences were just wrong because they got in the way of their recreation. People who would cluster by the river, on *his* land, fishing and scaring his cattle so they wouldn't move to another pasture. People who had no understanding about livestock. Slowly but methodically, agriculture was being pushed aside, turned into the idiot stepchild of the valley. Grazing permits on National Forest land started diminishing as ranches thinned out and the government reserved more of it for recreational use.

Still, the bikers and hikers weren't the worst. At least a lot of the recreation people went back to the cities they came from. But what about the ones who stayed? The ones who were buying land, buying those former ranches that had been taken over. "Thirty-fived" is how the locals disdainfully described it when damn fine ranchland was diminished, shaved into thirty-five-acre parcels with a pretty house, a horse or two, and some sculptured landscaping for privacy. And, hell, it was happening all over the valley; it seemed like there was no end to the people who were willing to ante up for a home site, no end to what they'd pay. It was almost like the dynamite once used in the mines had been put under property values and detonated because prices were sure exploding. Land that Sheldon Trampe had once paid $100 an acre for was going for $10,000 an acre and hurtling higher, each new wave of appreciation feeling like a loveless good-bye kiss to people like Bill Trampe.

Over two decades, the number of working ranches in the East River Valley went from fifteen to four. In their place came subdivisions like Hidden River Ranches, Roaring Judy Ranches, Glacier Lily Estates, all of them growing like weeds only you couldn't spray them. Now there were even these ridiculously large homes near the golf course—*the golf course!*—in Crested Butte. Trampe would come down from a day of herding cattle in the pastures, a day of "recouping" on the land, and he'd just shake his head when he passed those houses, many of them not even sold—just built on spec, just standing empty, almost like their silence was mocking him. Day after day he'd do this, riding an emotional roller-coaster that just made him sick to his stomach. And if that wasn't enough, what about the people who called the sheriff to complain because Trampe was making too much noise cutting his hay with a diesel-powered machine? What the hell was he supposed to use—scissors?

One day—although it wasn't as sudden as that—he turned around and realized he was outnumbered. Even the people in local government weren't natives. Oh, sure, they *thought* they were. But if you'd asked Bill Trampe about that, he'd have told you "People who have been here thirty years, they think of themselves as native residents. And we accept the fact that they think they're natives, but deep down inside I know they're not natives. But they're the people, frankly, who are running our communities anymore; they're our politicians, they're our county commissioners, they're our city councilmen . . . the natives, there aren't enough of us, and we feel rather pushed aside. Our philosophies and our ideas about our community don't match what the majority of the population wants. I don't make a big issue out of it, but . . . ." And then his voice would've trailed off and he wouldn't be smiling.

Hell, you can't *become* a native. You're born that way. Bill Trampe began to feel squeezed. He knew "exactly how the Indians felt. I'm the modern-day Indian and the cows are modern-day buffalo. When we're about to become extinct, they'll be an effort to save us, but I think it'll be too late."

And if he couldn't help but feel mistrustful and resentful of newcomers, they couldn't help but feel the same toward him, toward all the ag people. During one confrontation, Barb East, Trampe's girlfriend who worked as a range rider, herding other people's cattle, was accosted by a member of a homeowner's association who demanded, "When are you people going to learn you don't belong here?"

You people? Don't belong?

What about all the sweat his grandfather and father and mother—and he—had poured into the land? What about the stewardship he was trying to practice? What about the fact that all these newcomers seemed to be "preservationists," while he was a "conservationist"? Someone who wanted to make sure the land wasn't just pretty, but strong and vital. Someone who wanted to make sure those pastures were prodded and coaxed into productivity, make sure the cycles of growth and harvest and regeneration were respected and perpetuated. What about the fact that *those people* were screwing up the migratory patterns of animals when they thirty-fived their land with fences and created that ugly "sprawl"? Whose land did they think those animals would head for now? How the hell was Bill Trampe supposed to have enough hay to feed his cows when a herd of 1,000 hungry elk showed up on his land in winter and started devouring the crop he'd broken his back all spring, summer, and fall to harvest?

You people?

Didn't he already have enough to do in those winters when the snow was so deep and persistent he'd have to plow it off his pastures so newborn calves wouldn't freeze to death after just minutes of life? Didn't he have enough to do to fix the miles of fences mangled by those winter snows? Didn't he have enough to do trying to figure out how he was going to do all this when he could barely find anybody willing to work for him? Hell, he hadn't seen an Anglo come ask for a job in forever and a day. Ranching didn't pay enough and even if it did, nobody seemed to have the stomach for the work. Thank goodness for young guys from south of the border. Thank goodness for Marcos, his foreman. Solid. Hard working. Reliable. But there weren't a lot like Marcos out there. So Trampe would hire who he could and hope they'd last the season.

Of course, he didn't expect anybody to endure what he did. Didn't expect anybody to have a horse fall on him and break his ankle and still get up the next morning to feed the cows, hobbling around with a cast fastened to the busted joint. Hell, you work with livestock, you're going to get hurt. You just accept that. Besides, it wasn't like he was being heroic or anything when he limped around feeding the cows. Those animals needed to be fed and somebody had to do it. And when you're a rancher, that somebody is usually you. Something happens, you just deal with it.

So that's what he did. Before all the newcomers could circle their wagons around the valley and drive him and all the ag people out, he decided to do something.

For a rock-ribbed Republican and a guy *Time* once called a "crusty rancher," Bill Trampe hardly seemed the type to help found a grassroots organization like the Gunnison Ranchland Conservation Legacy. But there he was, around 1995, joining with ornithologist Susan Lohr to create the group that would keep some of the land from being thirty-fived. It wasn't easy. Ranchers are suspicious—hell, *Trampe* was suspicious. Even of Lohr, who arrived in Crested Butte from New Hampshire in 1986 to become director of the Rocky Mountain Biological Laboratory. Although he had to admire anyone who was willing to spend winters in the remote ghost town of Gothic, Trampe didn't fully trust her at first. Most scientists were strict environmentalists, people who didn't want to listen to ranchers.

But Lohr was different; she and Trampe had "the ability to communicate with each other." Still, it wasn't until they both served on the Upper Gunnison River Water Conservancy Board, which was fighting the perpetual battle to prevent the Front Range suburbs from snatching away West Slope water, that Trampe was able to dismantle the last wall of suspicion between them.

Once the bond was formed, the two personalities fused into an impressive force. Maybe Lohr wasn't a native, but she sure thought like one. She knew that hay and cattle were good for the land. She knew they were crucial ingredients in natural ecological cycles—and fancy housing developments weren't. Over the years, Trampe and Lohr nursed the Gunnison Ranchland Conservation Legacy along, talking to ranchers about conservation easements, letting them see that the easements were a way to preserve the land for agriculture, a way to ease crippling inheritance taxes and pass it along to heirs. A way to prevent what had happened to the Rozman property. At one time, when his son was just a small boy, Sheldon Trampe had grazed his cattle there. Now all that good pasture land was gone, turned into those Glacier Lily Estates, turned into *sites*, places where people "don't make a living," but "just play."

For his part, Bill Trampe was doing what he could with his own land. He persuaded his mother and sister—co-owners of the 6,000-acre ranch—to place an easement on 978 acres just north of Almont. He wished it could have been more, but that was a start, a show of his commitment to the land. The same commitment he had shown when AMAX had come courting back in 1981.

The enormous mining company was eyeing 800 acres of his land for a molybdenum mine tailings pond. Naturally, being international and rich, AMAX offered to buy the entire Trampe Family ranch just to get the acreage it wanted. Sure the money was tempting, but for Trampe "success is not financial." Sure he wanted to be comfortable, but his level of comfort was different than "some urbanite's."

Besides, what would he do without ranching? Yeah, the hours were hell and Mother Nature could make you insane, but from Trampe's perspective, "This is what I do. It's the only thing I know and I truly love doing it. I don't have a desire to do anything else." He'd spent a lifetime learning how to ranch—did anybody really think he was the type to just go off somewhere, count his money and *relax?* And it's not like he could just pick up and go ranch someplace else. Not when he'd spent his whole life being intimate with the pastures and the weather, with the *ways*, of the East River Valley.

Anyway, he had a better idea—trade those 800 acres the Trampes owned for another piece of land. Like, say, those 2,000 acres near Crested Butte, land that was slated for trophy homes and subdivisions, for the kind of development that would create more sprawl and fragment the habitat and not add one goddam thing to the valley but would sure take away a lot. AMAX didn't care about agriculture or development; AMAX cared about a tailings pond. AMAX said sure, bought the land from the developers and made the deal.

The deal made him feel pretty damn good. Just like he'd feel pretty good two decades later when the Gunnison Ranchland Conservation Legacy was cranked up and rolling. How could he not be pleased and proud? After thirteen years in the easement trenches, the Conservation Legacy had put together enough deals to save nearly 16,000 acres of land from development. Land that wouldn't be thirty-fived. Land on which nobody was going to build a restaurant where snooty people would look down at a man who had a little manure on his boots.

Bill Trampe is behind the wheel of his 2500 HD Duramax Diesel Chevy pickup pulling 11,000 pounds of cattle to the last big feed of their lives at a pasture near Crested Butte. The ten cows are all "dry," they didn't birth a calf this spring, didn't add to Trampe's herd of 800. And, what's more, they aren't going to get a second chance. Harsh? Maybe. But ranching's not a profession for the sentimental or the faint of heart.

Maybe that's why Trampe sounds more blunt than mushy as he talks about how agriculture became such a "minor deal" in the valley that the last machinery dealership here went bust. Now, when he needs a replacement part, he has to travel sixty-five miles over to Montrose. It's an inconvenience, but what are you gonna do? Even when he talks about something that ought to be more emotional, there's a slight edge to his voice. Like now, for instance, as the truck swings past the Elk Mountains, which are so beautiful and rugged they almost don't look real, and Trampe is remembering a time when life up here was different.

"If things got to you, a guy went to the hills for a day and you got away from what few people were here—we thought it was a lot of people then. You just went to the hills, and were totally isolated and recouped yourself," he says, eyes on the road. "And I still do that to some degree in the fall, but I never get totally away from people anymore. I have to get totally back in the wilderness where very few people go to get that sense of aloneness I need to recoup."

He makes that memory sound more matter of fact than you suspect it is. You suspect this because you hear him talking about the time four years ago, in October, when one of those early fall snows slammed the mountains and the "countryside just froze up." He had some cattle up there that were just too scared to come out along this one rocky trail because of all the ice and snow. He'd spotted them with his field glasses, wandering around up there at 12,000 feet. Knew there was no way he was gonna get a horse into that country. So it was just him and his dogs, hiking in, four, maybe five hours in the snow. It was dark by the time he got all the cattle out and years later he remembers, "What an exhilarating experience that was. Those are experiences you have once in your life. I don't want to ever do that again, but I'm glad I got to do it once, y'know?"

You do know. You know because there's something in his voice when he tells you about it. Something that doesn't have any crust on it at all.

Now he's talking about natives, people like him, people who look up at the trees on the mountains and don't see aspens but "quakies," who don't cross creeks but "cricks." He's talking about people like the guy who used to live in the house that he's driving past. That guy was a year behind Trampe in school. Died of pancreatic cancer a year ago. His son is in his mid-forties. Trampe knows this because that's what natives know. Just like he knows the people who live on the land he's now driving past. A mother and son. Old Italian family. Set in their ways. Good people. Ranchers, even if their herd is only fifty strong. People like him. People who know that "development and agriculture don't mix if you're a true ag man."

The truck crosses Ferris Creek, near the old Stubbs Place, where a homesteader once worked the land that some of Trampe's cattle now graze on. This is the land that he traded AMAX for and it is spectacular. The air at 9,000 feet is so clean it feels like it was scrubbed with a pine brush. Whetstone Mountain and Mount Crested Butte slash at the sky, and the grass runs green forever. True, there's a lot of purple larkspur—poison for cows—in full flower, but Trampe hopes the cows will stick to the healthy grass. He lets them out of the trailer and watches as they amble toward the smorgasbord of grass. He should have had the cows up here before the larkspur ever bloomed, but the damn snows of winter and spring put him behind, busting up the fences that define the pastures and forcing him to juggle his grazing schedules.

And after he juggled those grazing schedules, well, then he had to figure out how to tackle those busted fences. Nothing new about that. Repairing fences is one part of ranching that never ends. He's hoping his young crew from south of the border will be able to fix plenty of fence before they wear out from the hard work. He knows they'll wear out sooner or later because, unlike him, they have a choice.

You bet there are times when he gets up in the morning and he'd like to shout, "Boy, I feel crappy today; I'm not going to go to work," but he can't because that's "just not an option for me." You bet there are times he looks around and sees how far he's fallen behind because of the weather and common sense is begging him to say, "I give up. The hell with this year, we'll just start over next year." But pretty soon his heart will tell common sense to get lost because his heart knows "it doesn't work that way."

It doesn't work that way, but he sure wouldn't mind if it did. That way, maybe all those aching body parts of his could get some rest. Maybe that surgically rebuilt right ankle—the one that's been broke three times—would stop screaming with the arthritis that's "just inundated the whole damn joint." Hell, he's "just as miserable with pain as I was prior to the surgery." And what about the arthritis that lives in his shoulders? Or the disc problems in his back? Or the neck that hasn't been right since he broke it during one of his "crash and burns" on a horse?

You get him on a bad day when his body is just one big hurt, and you might hear the guy whose father and grandmother both died at sixty-three say, "Some days I don't care whether I make it to sixty-five. If the physical deterioration continues at this rate, I don't want to be around too much longer."

But most days it isn't like that. Most days, it's more like the hell with the pain because, "There's a lot of fire in me—that's what keeps me going. Physically, I know I can't do a lot of things I used to. Sure, on a one-day basis, I'll work any of these young guys under the carpet. I just can't do it day in and day out like I used to. I just got too many injuries and arthritis. There's just not enough gas left in the tank, physically. But mentally and emotionally? Yeah I love this. That's one of the things I'm fighting now—I know I gotta slow down physically, but I still want to do it, y'know what I'm saying?"

By now he's driven back south on County Road 135 and is unhitching the cattle trailer in the front yard of the house his father and grandfather built, the one where he lives now. He's hitching the flatbed trailer to the pickup because he has to head out and gather some hay for the yearling heifers he's keeping in a nearby corral. He stops to talk to Steve Guerrieri, a neighbor who's using Trampe's corral to run a "rectal palpate" on his bulls, to "see that there's no adhesions to the ejaculation organs inside, no ruptures. See, they take the semen and see how viable it is." Apparently, Steve is also testing for "trichomoniasis—that's a venereal disease spread by bulls." Trampe and his neighbor make friendly small talk although neither man has time for too much of it.

Trampe pulls the flatbed off the road at Jack's Cabin, where he has 300 acres of hay meadow. A long time ago, Jack's Cabin used to be a stop on the stagecoach route, a trading post for the trappers and miners who were the first white men to reside in the valley. Now, it's just a vague area with a name, a place whose history hardly anybody knows.

To the east are the Roaring Judy Ranches, pristine ranchettes splayed across a hillside that's been thirty-fived. Trampe doesn't give them a glance. Instead, he gets out and walks about a quarter-mile to a tractor, a slow, purposeful amble. Not the walk of a sixty-one-year-old man whose ankles are shot. Maybe he's got no time for the pain right now. He's busy. He comes back with the tractor and, manipulating the heavy machinery, begins moving the 700-pound bales that are stacked together onto the flatbed. "If I was a real man, I wouldn't need a tractor to load 'em, would I?" he says, laughing.

Sunlight glints off the pieces of straw that are blown from the bales. Soon, they become a small snowstorm of hay whirling in the wind that has come out of the north. Within twelve minutes Trampe has deposited fourteen bales onto the flatbed and is ready to bring it to his yearling heifers.

On the way back, he talks about this and that. About how he's sort of made his peace with the mountain bikers, how they stay off his land now and confrontations with trespassers are mostly in the past. Some ranchers actually let the bikers use their property for a fee, but not Trampe. "I don't want to go there," he says simply. Then, because recreation is a hot button with him, he has to add, "Look, the public needs to understand, this is private property and right now, it's primary use is not recreation. If you want to enjoy it from afar, welcome to it. I'm not asking anything of the public except to stay away from my business. Eighty percent of this damn Gunnison County is national forest ground and *that's* the public ground."

He talks about issues with the county government officials, how it seems they won't fix the fences that separate his land from county roads even if their snowplows "annihilate" those fences during the winter. He talks about the problems he's had with the county about cleaning the barrow pits, the ruts that run along the side of the road in front of his house and are used for drainage. You hear the click of another hot button, so you just sit back and listen:

"A year ago, I got the [irrigation] water going good and the barrow pits were full of crap, and the water was going over the road. Well, they [the county] got on my ass and wanted me to shut the ditch off because it was running all over the road, and I said 'Hell no—you maintain your damn road properly and give the water a place to go. Well, we had a big brouhaha, and I finally went to the county attorney and said, 'You guys have governmental immunity, and you're just setting me up for one helluva big wreck. Somebody's gonna come flying down that road and hit that water and roll their vehicle and kill themselves and they're gonna look for somebody to sue.'

"And I told them, 'The barrow pits are full of crud and there's no place for the water to go.' And I said, 'I know what you're trying to do, trying to force me to give you right of way on my property and I won't do that.' I said, 'Well, let's get with it or fix the road.' Well, they stumbled around and thought about it all winter, and finally they got out here to clean the pits. There they are now."

By the time he unloads the hay, it's past noon. You offer to buy him lunch in town, but he doesn't have the time or patience to sit in a restaurant. This afternoon has been sort of penciled in to clear irrigation ditches, "play in the water," he says with a smile. That's the afternoon job. The job right now is "to find a sandwich."

He takes you on a quick tour of his unpretentious house, the one he says "isn't fancy." It is sparsely furnished, but comfortable. A lot of his girlfriend Barb's artwork adorns the walls, but, except for the watercolors and sketches, it looks like the home of a man who's been a bachelor all his life—and who intends to stay that way.

It doesn't take long to construct sandwiches of ham and cheese, with some potato chips on the side. The morning's coffee still tastes pretty damn good. In between bites, he talks about his mother, who is still alive and living in town, how she's a "difficult woman," but how she's "strong-willed and that's why the ranch has survived." He talks about his sister, Bette Mae, and how he's not so sure she's on the same page with him about easements and keeping the land they own jointly free from any development. He talks about how it felt to testify before the U.S. Congress on irrigation-ditch right of ways on federal land a few years ago, how he's been on a number of water conservancy boards and how "water is a real passion of mine." He talks about how when men become politicians "the world changes them; they're not the same," and how he's known a few men like that and his voice tells you the way those men have changed doesn't exactly please him. He talks about Barb and her art and how she's not a valley native but she's sure the next-best thing. He talks about his father and grandfather. He talks about growing up and all the changes he's seen. He talks because he likes to talk and because he knows you've come to hear his story and, truth be told, he doesn't seem to mind sharing that story. Hell, he knows it's not boring.

As he tells more of that story, you wonder what all those mountain bikers and hikers he's exchanged curses with, what all those newcomers who've complained about his loud machinery next to the land they've thirty-fived would think if they could just visit with Bill Trampe in his old-fashioned, sparse kitchen and maybe share a ham sandwich and drink some pretty good coffee with him. And then you think how that is never going to happen.

At least the screaming matches are a thing of the past. At least nowadays if he encounters hikers on national forest land and he's moving some cattle and they get all pissed off he's more than likely to let them rant and say nothing. Just sit back on his horse and laugh some because it keeps him calm even if it maybe does the opposite to them. "Kinda crusty, huh?" he says, smiling as if he doesn't mind being that way at all.

Then lunch is over and it's time for him to go play in the water. He gets up from the table, lifting the frame that's inundated with arthritis in the ankles and shoulders and hurts almost as bad in most other places, and heads out the door, ready to work the land. Ready to do his part to keep the world he knows and loves from being thirty-fived. Ready to earn a living the only—and the best—damn way he knows how. Ready to do the things that can bend a man's body but straighten his soul.

Wetlands below Mount Crested Butte

IRBY RANCH

- *Gunnison County*
- *Owners: Bob and Irene Irby, Stan and Bonnie Irby, Dale and Wendy Irby*
- *First ranched in the 1870s*
- *Conservation easement held by Colorado Open Lands*

Irby Ranch lies in the Gunnison River basin along Tomichi Creek, the eastern fork of the Gunnison River that meets the main fork near the town of Gunnison. Tomichi Creek is remarkable for the manner in which it meanders through the valley: curve after curve of slow water bisect some of Colorado's most spectacular hay meadows, all within sight of U.S. Highway 50. The Cochetopa Hills provide a fine backdrop to the south, and the Sawatch Range to the east is visible from most of the ranch. This valley is quite special in August when large, round hay bales decorate the meadows.

First to ranch the Irby tract was the Snyder family, who did so despite warnings from resident Ute Indians that the winters were unmanageable—which proved true—and they moved away in due course. Vernard Irby migrated from Texas to the area by himself at age fourteen and worked odd jobs, including mining and cattle tending, then bought the ranch in 1942. His son Bob and grandson Stan Irby operate the ranch today.

Chipeta Mountain

Tomichi Creek

COGAN RANCH

- *Chaffee County*
- *Owners: Cogan family*
- *First ranched in 1872*
- *Conservation easement held by Land Trust of the Upper Arkansas*

The original Cogan Ranch lies along the banks of the upper Arkansas River, four miles south of the town of Buena Vista. Another parcel sits to the east on top of Trout Creek Pass. Though the views of Buffalo Peaks from that place are remarkable, it's the pastoral hay meadows along the river that define the ranch. Irish-green grasses fill the valley, and from one particular vantage along U.S. Highway 285, the Arkansas seems infinite as it bisects the ranch on its way south to the town of Salida. Cottonwood trees, lime-green in May and orange in the fall, decorate the banks. Billy Huey homesteaded some of this land in 1872, then sold a portion of it to Irish immigrant Jeremiah Cogan in 1892. Cogan's sons John and William continued ranching and farming the land, and today grandson Joe and great-grandson Bruce operate the ranch.

Arkansas River

Mount Princeton sunrise

Arkansas River looking south in fall

Arkansas River looking south in spring

Buffalo Peaks

ROCKY MOUNTAINS

RUSK HEREFORD RANCH

- *Custer County*
- *Owners: Rusk family*
- *First ranched in 1872*
- *Conservation easement held by Colorado Cattlemen's Agricultural Land Trust*

There may not be a more beautiful mountain valley in the American West than the Wet Mountain Valley, which is the location of the Rusk Ranch. Immediately to the west lies the Sangre de Cristo Range, to the east the Wet Mountains. The Sangres contain several fourteen-thousand-foot peaks, which some years remain snowcapped in August, and extend as far to the north and south as one can see from the ranch. Though not as stately, the Wet Mountains do their part by enclosing this little valley and protecting its ranches from intrusion by outsiders—the valley is isolated and pretty much at "the end of the road." Though subdevelopment has dissected much of the outlying foothills, the valley floor remains essentially intact as working ranchland.

Harvey Rusk worked haying jobs in the valley in 1936 and settled there in 1946 on land owned by his father. He then proceeded to teach himself the business of ranching. Today his son Randy and grandson Tate continue the tradition on the original ranch as well as on the contiguous Beckwith and Kennicott ranches, both ranched since the 1860s, that they lease from their owners.

Kennicott Ranch

The man who lives just this side of Never Never Land is sitting in the best restaurant you never heard of, talking about a lot of things, including the fact that one of the reasons he hates big cities is because "The air has all been breathed." He doesn't smile as he says this because he's serious. But then, he's serious about most things that have to do with elbow room, sweeping vistas, and the Wet Mountain Valley, the place where he grew up, the place he returned to, the place whose solemn beauty and free-rolling *space* mean so much to him he was willing to risk his money, his good name, and friendships he held dear in order to do right by it.

His sacrifice was prodded by a sense of responsibility and love, by a stubborn integrity and an even more stubborn pride. A good kind of pride, the kind that makes a man care about the quality of his cattle even if they don't bring the best price because the market's gone to hell. The kind of pride that drives him to build a special shoe for a mare with a misshapen hoof when a lot of people might have said screw this and gotten rid of her. The kind of pride in ranching traditions that prompts a man to buy the rights to the brand of Frank Kennicott, which was registered before Colorado ever became a state. The kind of pride that has him devoted to his land and his calling even though he knows that "If you want to make a small fortune in the cattle business, you better start out with a large fortune."

The man's name is Randy Rusk, a strong name, almost an abrupt name. A good name for an old-time town sheriff, which he sure looks like. Dark, brooding eyes with both an edge and a twinkle. Clipped salt-and-pepper moustache. Face a bit rawboned, cheeks chapped by the wind and burnished by the sun. Broad shoulders, strong hands. And all of it fashioned onto a six-foot-three-inch chassis that looks bigger than that. He even sounds like a sheriff from the Old West: low, quiet voice coiling out of lips that hardly seem to move. A no-nonsense voice that deals in a lot of common sense and not many frills and frequently gathers momentum like a stone rolling down a hill.

Seated with him at the table inside Yoder's Restaurant in Westcliffe—where a ham-and-cheese sandwich tastes like a little bit of heaven—is his wife Claricy. They've been married nearly forty years although, truth be told, they've known each other way longer. Claricy figures "We met in the sandbox. It seems like I never didn't know him." She has borne Randy three children—first Mysty, then Lisa, and finally Tate—and is as much a sheriff as him. No, she doesn't have that same watch-your-step-glare that he does, but that's about all she lacks. See, she's been forged by the weather and the agricultural way just like Randy. They work together on their ranch, tending cattle, dragging the meadows ("turd busting," she explains) on their mammoth tractor, moving irrigation dams, keeping the books, stretching wire from fence post to fence post, you name it.

"Ranching is just what we do, what we love to do, what we *know*," she'll tell you. Then, "Just look at my hands—I don't have the hands to work in a bank or a jewelry store. They're just too rough."

She says this with a laugh and without regret. She's proud to be a rancher, wouldn't trade her life for anything, wouldn't trade the land for anything. She knows what happens to your soul when the pastures and their asymmetrical beauty disappear. In stray moments—even ones with strangers—an old pain springs up and drags a few tears from her eyes. It's the pain of land that was lost, land that her father had to sell to save his life, land that doesn't grow grass and feed cattle, land that's been chopped up and homogenized into neat little parcels with pristine "ranchettes." Land that's had the spirit choked out of it. Land where you *can* have the kind of hands to work in a jewelry store.

But the tears don't last long—and neither does lunch. Too much work back home. So the two sheriffs pay the tab, climb into their pickup, and head back to the ranch where Randy was raised, where his mother and father still live. The ranch that sits just this side of Never Never Land in a magical valley worth fighting for.

Twilight is starting to slide over the Wet Mountain Valley as Randy and Claricy head out to change the positions of some irrigation dams. Thick sheets of orange-colored plastic are turned and shifted, diverting the water, sending it slithering in new directions. Tomorrow, to comply with local covenants, they'll have to relinquish their shared water rights to other neighbors for ten days, so right now it's important they maximize the flow across their land.

Of course, tomorrow will find other tasks bubbling to the surface. "We're in the frenzy season now," says Claricy, referring to the transition from calving to branding to irrigation. Soon enough, there will be the transition to haying. Then something else because, well, there's always something else. Even in winter, when things slow down, they don't stop. Watch the herd; if the snow is deep, get your butt out there and make sure they have food. Repair that patch of fence that some cattle ran through. Or maybe see about that tool that hasn't worked properly for God knows how long and you swore you'd fix once you found the time.

Meanwhile, as all this is orbiting around your skull, you better realize that you might wind up doing something unexpected. As Randy says, "Very seldom can we stay on course. There's no set schedule, but there is a plan." He might be smiling as he says this, but you can't be sure.

One thing that is for sure right now is he's no longer talking. Instead, he's zipping along the ranch on his ATV. Trailing behind is Claricy, driving a little slower, a little more tentatively, like someone who not a year ago wiped out on her ATV and ruptured her spleen. Funny, though, as much as that hurt, it was nothing compared to that terrible pain she's been carrying for thirty years, the pain that clawed a hole in her heart, climbed in, and just won't leave.

When Leslie Schulze found out he had emphysema and had to get out of the Valley and down to a lower altitude, he didn't have a whole lot of time to make arrangements. He knew that none of his five kids could afford to buy his land, so he sold it. Nothing else to be done. The ranch where Claricy had grown up—where she'd learned to ride and wrangle and do all the nitty-gritty things that define a cowpuncher—became a ghost that haunted her. As the land got subdivided and turned into those neat little habitats for city people moving into the Valley, she grew sadder. Over the years, she couldn't go near the property, couldn't even look at it. It made her sick.

She wasn't feeling too much better in the late 1970s, when those developers came and built Conquistador, an ill-fated ski area, in their midst. She and Randy called it the "Scar," and tried hard not to look at what it had done to part of the Sangre de Cristos. They tried not to think of the unimaginable—that their beloved valley, where they had grown up, where good green grass and ambling cattle once held sway, was being invaded and taken over.

In Westcliffe, where they had gone to the same tiny school, real estate offices were popping up like dandelions. Suddenly, it seemed—although the change had really been slow and insidious—Custer County was exploding with people. Not ranching people, but people who were suspicious of ranchers. People who would howl at the county to plow the snow from the roads that went by their houses and then, after those plows had accidentally taken down a stretch of some rancher's fence and his cattle wandered out onto those roads, would howl about cows being in their way. People who loved the scenery but didn't respect the land.

People who had no idea what makes Harvey Rusk tick.

He may have been the son of a doctor, but the gangly boy from the city knew that the cure for whatever ailed him was being up on a horse and riding. That's why, back in the summer of 1936, when he was not quite sixteen and still wet behind the ears, Harvey Rusk rode up from P-eblo and into the Wet Mountain Valley. He and a buddy camped out, looking at the stars at night, then getting up to work haying jobs for a dollar a day. Just had a helluva good time. So good, Harvey kept coming back. Kept coming back because it didn't take him long to realize that "I'd just as soon be a cowboy as anything else." Didn't know why he liked horses so much. Didn't know why he liked cattle so much. Just did.

Neither college nor the army changed his mind about a career, so when he mustered out of the service in 1946, he took his bride of three years and settled down on some land his father had bought. What he knew about ranching you could squeeze into an eyedropper, so he hired a fifty-year-old ranch hand to help him out. Man's heart went bad and he quit after six months. Harvey saw his first calf born all by himself. *Damn*.

Mistakes? He made them by the barrel. Didn't matter; he learned. Some of the older ranchers would come by and chuckle, maybe even snicker at Harvey's screw-ups. Didn't matter. He was learning. Kept right on learning, too. He learned so well that more than sixty years after he and Jean settled in the Valley and raised their family, Harvey was still there. And all those fellas who snickered? To hell with them. Harvey had earned his spurs and was damn proud of that.

Harvey and Jean had three children. First came Harvey Dean, last came Susan Elizabeth, and sandwiched in between was Randall Conrad. Randy was most like Harvey—at least when it came to ranching. Just like his old man, he knew early on what he wanted to do. Like his old man, he went to college, graduating from Colorado State University with a degree in animal sciences. Like his old man, he married young—he wasn't twenty when he and Claricy wed. "I guess we were still pretty much in the sandbox when we got married, too," she would remember nearly forty years later.

After Randy finished college, he and Claricy—and baby Mysty, who was born the week of graduation—left Colorado. Randy started an artificial insemination business that led the newlyweds to Montana and Wyoming. Then he got a ranching job in southern Arizona, where Lisa and then Tate were born. Then it was out to California where he ran an embryo transfer company. But no matter where they went, the boy and girl who met in the sandbox always knew where they wanted to be.

Eleven years after they left, Randy and Claricy came back. They came back because the Valley's pull was too great to resist. They came back because they wanted their kids to grow up where they had. Wanted them to attend the same school, breathe the air that wasn't already used up the way it is in a city. Wanted them to know what it was like to experience the different pageants of light as sunrise lit the Valley and sunset dimmed it. Wanted them to come of age with the Sangre de Cristos to the west and the Wet Mountains to the east, each range standing sentinel in its own way. Wanted them to feel wind so glorious and ferocious it could grab you and throw you around like a feather in a blender. Wanted them to know the sweet, rugged pleasure of exploring the other side of the Sangres, of going to a place that Randy had always known as Never Never Land, a place where big canyons melted into one another and you could ride for days if you wanted.

They came back and they worked hard because ranching, as Randy knew, "is damn hard work." They skirted on the edge of profit and loss because, sooner or later, that's what happens to most who love the land. They'd try to soften the hard numbers of ranch economics by running hunting trips up into the mountains. They sold some of the hay they grew. They set up a game processing business right next to their house, the one Claricy called "an old metal barn;" the one Randy said, "is not exactly the romantic Ponderosa." They juggled and they struggled and they worried. They stayed "bull-headed, stubborn, and independent and probably not very logical some of the time" because that's what most ranchers do to survive.

And then survival became more than a shade easier when providence ambled by and tapped Randy on the shoulder.

It happened the summer of 1984. A camera crew from a big city ad agency came to town, working on a commercial for Busch Beer. Nobody in the Valley had ever heard of Busch Beer, but who cared? The crew was looking for a few good cowpunchers and after they held tryouts, it didn't take long for them to see that Randy was one of the best. They had him join the Screen Actors Guild and then spent the next two days filming him and some other Valley guys "doing this, that, and whatnot," riding and roping, doing, y'know, cowboy stuff. The whole thing was "weird enough to be fun" and after the crew left, Randy figured he'd probably earned a hundred bucks a day, good dough for a cowpuncher, especially for doing what came natural. Then the money arrived and he found he was dead wrong. In the first place, the agency paid for full medical and dental insurance for him and his family for two years. On top of this bonanza, whenever the sixty-second commercial was aired on TV before some big football game—and we're talking years and years—"we knew a check was coming." A residual check. And although he won't say how much, it's likely the checks were more than a hundred dollars a pop.

When he bothered to think about the whole commercial thing, it didn't make sense. It was the kind of experience that could make a man whose feet never left the ground shake his head, sort of smile, and say, "You stay up at night, wondering how you're going to make a profit, and end up losing your butt a lot of the time anyway. And then this thing falls out of the sky."

Well, maybe that was weird, but, when you think about it, probably not as weird as "making million-dollar deals on a handshake." Which is what Randy—and most other ranchers—usually did. You gave a man your hand, you gave him your word, you'd better follow through. Oh, sure, over the years, things would change and get more complicated. Like you'd sit down to do some deals and there'd be "these two law dogs in the middle and all these papers floating around."

But some things didn't change. Like, let's say you showed up at a man's ranch wanting to buy a horse he had, but only had half the money he was asking. Let's say you told him you'd have the other half after you sold your calves. Chances are you'd ride off on the horse that very afternoon. That's the way things were done. You said you'd do something, you did it. Simple. A pride thing. That was the way Harvey was brought up. The way he taught Randy to be. The way of the Valley.

But the Valley was changing.

Something was going haywire—Randy and Claricy and Harvey knew that. They knew that the land wasn't supposed to be subdivided and subdued; it was supposed to be cultivated so it would give back. They knew that, as Randy would say, ranchers "live in a world that is more real, more basic" than the one inhabited by most newcomers to the Valley. They knew that ranchers live closer to the truths of the earth. That "Ranchers are the most pure conservationists there are. Our existence is a reflection of being able to take care of the land, asking the land to reproduce for us and reproduce as much as it can." They knew that the land was "more of a responsibility than an asset."

Ranchers helped feed the country, the planet. And not just feed it, but feed it "with some of the safest food in the world—and the cheapest." It was that "pride thing" again. The kind of pride you developed when you were young and never let go of for any reason—and certainly not because somebody came by and started waving money in your face if only you'd put down your dream the way you'd put down a horse with a broken leg. The kind of pride that made it a sin to yield your land to people who only wanted to use it as some kind of prop. The kind of pride you felt when you made a little bit of money and you knew you were doing what the earth wanted you to do—"take sunshine, water, and labor and help it produce." You didn't just let the land alone because "No use is abuse, if you ask me." Problem was, as time went by, few people seemed to be asking.

Fewer still seemed to care. About the land or the spirit of the Valley. Things were changing big time. Randy remembered when county sheriff used to be a part-time job. Now there were something like nine full-time deputies. He remembered when it seemed like you knew everybody's name in Westcliffe—maybe even in Custer County. Now the faces didn't even look familiar.

But that wasn't the real problem. The real problem was that the *Valley* didn't look all that familiar. Twenty years ago, it was "cluttered" with big, working ranches, and now there were maybe half a dozen. When Randy and Claricy had left in 1972, there were probably 20,000 cows in the Valley, and now "you'd be hard-pressed to find 2,500." Agriculture seemed to be vanishing and to all those developers and real estate people who were catering to all those new arrivals that was just fine—hell, it was preferable. To Randy it was something else. He looked around his property and saw himself surrounded by subdivisions. A lot of days it probably felt like all the air had been breathed.

Okay, *enough*.

Randy began studying the concept of conservation easements. At first he thought they were about putting roads on your land. He soon found out otherwise. Easements could protect the land from developers. Easements could keep the land producing. The catch was that the land's worth would decline with an easement—at least on paper. He spoke with Harvey. His father's response was "I'd rather see cows on this land than houses. I want this to be cowboy land."

In 2002 the Rusk Family sold development rights for its 1,500 acres to a public trust, ensuring there would never be a subdivision built on it. Right there, in a heartbeat, Randy figured his net worth plummeted "a third to a half." Didn't care. He then bought out his brother and sister's share of the property for a chunk less than they would have made selling to developers. Although Harvey was on board—and Harvey called the shots pretty much—tensions flared within the family. Harvey Dean, Randy's older brother who owned a dude ranch in the Valley, was less than thrilled with the easement deal that had devalued the land. Did that cause bad blood? "Well, it didn't make any good blood." End of conversation.

But it wasn't just family who got "torqued." Friends, neighbors, people Randy had known forever, turned their backs on him. *You burned your own barn down, Randy.* They were angry that he might be throwing gasoline and lit matches their way, too; angry that he might have hurt their own potential for profit. After all, as agricultural property, Valley land was worth about $50 an acre. As potential sites for subdivisions, it was worth $7,000 an acre.

Randy found out just how molten the resentment was when he ran for county commissioner in 2004. An ugly campaign ensued. His opponents—who sometimes seemed more like enemies—claimed that if he was elected, well, there'd never be another house built in the county—not one. He was dangerous. He was the devil in spurs holding back the wheel of progress. Him winning would be the end of the world—or, at least, the Valley.

For Randy—"a cowpuncher, not a politician"—running for office was some kind of eye-opener. He found out that going through the political process was like "taking my pants down and letting everybody have a look." He found out what it felt like to be vilified and misrepresented to the point where "I got everything but physical beatings." He also found out how it felt to lose because that's exactly what happened come Election Day. Secretly, he was probably glad to lose the office; he'd run "more out of a responsibility than any kind of quest. It was mostly because a friend asked me to." But he wasn't so glad to lose a bunch of friends—which is also what happened.

Still, after things had settled down and he had time to sift through what he'd said and done, he decided he wouldn't have changed a thing. And as for the way the developers and the real estate people—some of whom had been friends—had portrayed him, well, that was a no-brainer. *I could give a crap.*

One thing you can bet he does give a crap about is how some landowners didn't see him as the devil. In the six years since he sold his easement on those 1,500 acres of Rusk property, a tide of conservation has flowed across the land, a tide that now protects more than 20,000 glorious acres in the Wet Mountain Valley. Randy knows this in his head and he knows this in his heart. He knows this because he helped set that tide in motion. He knows this because he and Claricy and their son Tate are leasing nearly 7,000 acres of that useful property themselves. Working it the way the land ought to be worked.

If Randy Rusk wasn't exactly sanctified in the Valley, at least he was celebrated outside it. In 2004, he was honored with the Aldo Leopold Conservation Award by the Sand County Foundation, one of the most respected conservation organizations in the country. Two years later, *Newsweek* honored him with one of its fifteen "Giving Back" awards, recognition for people who "use fame, fortune, heart and soul to help others." Rusk's photo—standing with Tate and Harvey—appeared in the July 10, 2006 issue, under the heading "The Pioneer." He was in pretty fancy company, too: Brad Pitt was a fellow honoree.

The awards are nice, but they're not what really matter. No, what matters are those 20,000-plus acres of land saved from being subdivided. Not one more house will be built on the Beckwith Ranch, the social capital of the Valley a century ago. Not one more house will be built on Frank Kennicott's spread, land where Tate and his wife Wendy and their kids now live and ranch the right way; land that was owned by a man who, says Randy, was "a real crackerjack," a man who not only pioneered personal branding in Colorado, but who started a local cattlemen's association and had progressive ideas about dams and irrigation ditches. A man whose legacy was important enough that being able to buy his FK brand was an "honor." No, not one more house will be built on any of this protected land; the march of ranchettes through at least part of the Valley and the San Isabel National Forest has been stopped.

So what if, in the end, all this saved acreage only amounts to maybe two percent of the Valley? It'll still be the most beautiful two percent there is; land that looks as wild and free as it did when a gangly sixteen-year-old kid came riding up on horseback from P-eblo and decided all he ever wanted to be was a cowboy.

Randy Rusk is busy. No news flash here because he's always busy. Even if he's not busy, he's thinking he should be. One time, at a family reunion, he got antsy because it was haying season and the weather was starting to look a little dicey and he figured it'd be a damn sight better to beat the rain and bale some hay than stand around and socialize. So he left before the family portrait was taken. Claricy had to have him photoshopped in.

Haying is a couple of months away, and he doesn't have to get the cull cows—the ones who can't reproduce—ready for their trip to the sale barn in Salida until tomorrow. So keeping busy right now means dealing with Peppalena. There she is, in the barn, standing patiently on three hooves while Randy works on her fourth. When she was about a year old, she cut her foot bad, so bad it became "more than slightly deformed" and, as such, she's higher maintenance than your average eight-year-old mare.

Before he nails on the special shoe he's built for her, Randy has to file down the misshapen left front hoof. Before he starts filing, he puts on a pair of chinks—shorter chaps—over his jeans.

"This file will put a chewin' on my pants, and I'll get a scoldin' if I come home with a big hole in my new pants," he explains to a stranger, sounding for a minute more like he's eight than fifty-eight. It's a sweet moment, but the stranger isn't about to tell that to the sheriff. Instead, he just nods.

Between bending the clips of the custom-made shoe around Peppalena's malformed hoof and then hammering it into place, Randy says through a mouth full of nails, "A lot of my friends said to can her, but, you know her mother, you knew her papa, so when they said there was a forty percent chance she'd be okay, you just sort of upgrade it in your mind to sixty and go do it."

He finishes nailing in the shoe, taking his time, making sure it's done right because shoeing a horse is also a pride thing. Then he looks up and responds to a question.

"Yeah, I guess doing this is uncommon, but she's a good mare. My wife likes to ride her. So I keep her."

Randy knows about the bond that exists between cowpunchers and their horses, a bond almost as strong as the one between cowpunchers and their land. He knows how important horses are to Heaven's plan because, "It's like a friend of mine says, 'The Good

Lord put cows on earth to train horses.'" He might be smiling when he says this, but he's serious as sin when he adds, "There's no connection like you get to a horse."

He looks away for a minute. "It's like being in a car when someone else is driving—I hate to drive a car. I'd rather be a passenger looking out the windows at scenery, at fences, at anything. With a good horse, that's how it is. You can look around. You don't have to look ahead of you, you don't have to push in the clutch or turn the steering wheel. A good horse'll just go."

Something moves across his eyes and he says, "When I can't work cattle on horseback, there'll be a big hole in my connection to the cattle, the land, the industry. A lot of magic will be gone."

Harvey sure knows what his son means. He's eighty-seven now and maybe he moves a little slower on those bad knees and bowlegs, but so what? He still gets up in the morning and checks the snowpack on his computer. Still gets out and moves irrigation dams when he has to. Still drives out to lend a hand when a few cattle wander through some busted fence. Still gets on the tractor, still paints, still repairs, still tinkers. After all, he's still a rancher, still a cowboy, dammit.

But in some ways he's a sad cowboy. A while back, Randy had to take his horse away. She was just too flighty; there was just too much chance of an accident. Randy brought him another horse, a calmer horse. Harvey said no. It wasn't *his* horse. Didn't want to ride it. Another one of those pride things. So he doesn't ride at all now, even though he misses it terribly. You can tell he does because sometimes he goes down to the basement of his house where he keeps his saddle and quietly sits on it.

Harvey hates not being able to ride. But he sure likes knowing that his family can. That they love to ride and ranch as much as him. Harvey likes knowing that he's kept this land as cowboy land, and that Randy will be there to keep on ranching, keep on *using* the land after he's turned to dust. Just like Randy and Claricy like knowing that Tate will be there to keep on ranching after they've turned to dust. Just like they're hoping that Tate's kids, their grandchildren—and maybe someday their great grandchildren—will be ranchers, too, will learn to love the traditions and the land as fiercely as they do. But maybe what they love most of all is knowing that whatever else happens, no Rusk will have to live in a place where the air has all been breathed.

Beckwith Ranch

Randy Rusk below the Sangre de Cristo Range

Beckwith Ranch

Frank Kennicott cabin, built circa 1869

CROSS ARROW RANCH

- *Conejos County*
- *Owners: Willett family*
- *First ranched in 1884*
- *Conservation easement held by Rio Grande Headwaters Land Trust*

Cross Arrow Ranch lies fifteen miles south of the town of Alamosa on the southern end of the San Luis Valley, one of the most productive farming and ranching regions in the state—and one of the most scenic. The Sangre de Cristo Range looms to the east and north, and the southern San Juan Mountains contain the valley to the west. Colorado's fourth-highest mountain, 14,345-foot Blanca Peak, soars above the broad, flat meadows of the Cross Arrow Ranch, looking remarkably similar to Mount Kilimanjaro above the plains of the African Serengeti. However, what might be even more impressive is the confluence of the Conejos River with the Rio Grande on the ranch's northeast corner. Both rivers are lined with cottonwood trees and wind through and along the boundaries of the ranch. In spring, their waters flood-irrigate the hay meadows, creating transitory wetlands not unlike seasonal wetlands in Africa. Parts of the Cross Arrow Ranch were homesteaded by Hispanic people as early as 1884. The Willett family acquired the existing ranch in 1971.

Cross Arrow Ranch headquarters

Confluence of Conejos River (entering right) and Rio Grande

San Luis Hills at sunrise

Rio Grande at sunset

Blanca Peak, the "Colorado Kilimanjaro," looms over the plain of the "San Luis Valley Serengeti"

Below: *Rio Grande and Blanca Peak*

EL RANCHO SALAZAR

- *Conejos County*
- *Owners: Salazar family*
- *First ranched in 1857*

The Salazar Ranch consists of several parcels of farm and ranch land on the opposite side of the San Luis Hills—the only significant escarpment in the southern San Luis Valley—from Cross Arrow Ranch. The original family homestead lies tucked up against the stark San Luis Hills in what is called Los Rincones—the corners. To this backdrop add the cottonwood tree-lined San Antonio River and views to the west of the San Juan Mountains, and one can understand why, at least as a function of the scenery, the Salazar family is still here.

Filipe Cantu settled here in the 1850s and married a Salazar. El Rancho Salazar was born when his daughter Maria also married a Salazar. The ranch has been in continuous operation under the Salazar name since 1888, and it still uses original Cantu water rights dating to 1857. Today, Cantu's great-grandchildren, led by LeRoy Salazar, raise beef cows on the original homestead. They farm potatoes and various grains on land acquired to the south of the homestead.

Potato storage

Sunset is approaching and the shadows are lengthening across Los Cerritos Cemetery, sliding over the wide slope of grass and headstones, falling on the only two living souls around. One is a stranger to the Valley. The other is a Native Son, a coiled spring of a man crackling with great impatience and greater energy; an intense man, a restless man, a man who hates waiting at red lights, a man who relaxes poorly and would almost always rather be working, even now, on the edge of twilight. But tonight he is making an exception because this visit is about blood and heritage, and nothing—not even the land—is more important to him.

He conducts a brief tour, pointing out the resting places of five generations of his family. Over there is his great grandfather, who came to Colorado before there was a Colorado. Over here is his father, who demanded much but gave more. Nearby is his older brother, whose death sixteen years ago is still fresh and painful. Not far from his brother are his two nephews, stillborn tragedies, angels gone before they ever spread their wings, proof of what his father used to say: *Dios da, Dios quita*—God gives, God takes away.

The Native Son knows this is true. He knows this just as surely as all those generations before him knew it. You live and work in the Valley—where the rain never falls enough in the burning summers, and the snow often falls too much in the icebox winters; where merciless hail can destroy a crop of beautiful wheat on the very day it is to be harvested; where the flailing wind comes out of nowhere, chases itself in vast corkscrews and sweeping circles that wreak havoc with your crop dusting and irrigation and then disappears back into nowhere—yes, you live and work in the Valley and you know all about what God can take away.

But if you are a Native Son, if you are LeRoy Salazar, you also know about what God gives.

Earlier in the day, before the cemetery tour, he had stopped his small, tidy, green pickup truck in one of his potato fields and looked out at the land that ran away in all directions.

"I go to church every Sunday, but I feel like I'm in church all the time—this is God's church, this great outdoors," he told his visitor, his arm gesturing towards the immensity of the San Luis Valley. "I mean, look at those beautiful mountains in front of us," he enthused, glancing at the Sangre de Cristos, words starting to tumble and run into each other, making perfect thoughts from imperfect sentences. "Look at that beautiful Mount Blanca over there, I mean, the thing about this, this place has an openness; it has, it gives me a sense of wonder that God can create things like these beautiful mountains.

"There is so much of wonder that every time I go out in the field and look up there and see the evening sun, and see the meadow and the forest, and the animals, and I hear all those wonderful sounds that you hear in the evening on the farm, whether it's from coyotes howling or owls hooting or from frogs ribeting. And then as night gets darker, you see these beautiful stars, and the Milky Way, and just this sky full of a sight that you cannot see in the city. It's just majestic here, it's a real marvel."

He paused, his face assuming an expression that bordered on beatific. The expression lingered, but the pause didn't. There was something else he had to say, something to finish the thought because LeRoy Salazar doesn't like to leave things unfinished.

"You know, to me, the Valley, it's almost like paradise. I'll be happy if paradise is like this, if heaven is like this. Wouldn't you?"

Then, quickly, almost as if he had just come out of a trance, LeRoy Salazar had the truck in gear and was speeding off toward his next task. Paradise would have to wait because there was much work to be done. Fields to be tended—barley, alfalfa, potatoes, canola. Cattle—Angus and Limousin—to be looked after. Hay to be moved. Irrigation systems to be turned on and monitored. Weeds to be dug up or sprayed, cell phone calls to be orchestrated. And, most important of all, there was his duty as an oldest son—to look in on his mother, to share the evening blessing with her, to appreciate one more day of life with her in the home where he came of age.

Land and blood, work and religion, crops and cattle, ranch and family—the themes swirl through the life of LeRoy Salazar. Indeed, they swirl through the lives of all the Salazar siblings—John, Ken, Elaine, Margaret, Elliott, June—as well their children and perhaps even their grandchildren. The link is umbilical, genetic, a complex strand of some immutable DNA. It runs across centuries and through all the Salazars, never to be broken, forever connecting the living and the dead to each other, forever connecting them to the Valley that has no beginning or end.

Imagine you are squinting into the light of a kerosene lamp in an adobe house as the stars tumble across the night sky. Imagine you are traveling back in time, far enough that you are sailing across the Atlantic Ocean from 16th Century Spain with Capitan Juan de Salazar, a man who will help found the city of Santa Fe. Imagine you can see the roots of the Salazar family tree take hold in the soil of the New World. From those roots emerges a young Mexican sheepherder named Felipe Cantu. He is just sixteen when a roaming band of Indians kidnap him. After several months, Cantu falls ill and becomes more burden than novelty to his captors. They give him to a Spanish trader from Santa Fe. When the trader dies, Cantu wanders north. By the late 1850s, he has helped create *Los Rincones*—The Corners—the first settlement in what will become the state of Colorado two decades later. Felipe farms the land, originally fourteen acres, which is fed by the San Antonio and Conejos rivers and is part of the great tableau that is the San Luis Valley, more than 8,000 square miles of land barely contained by the San Juan, La Garita, and Sangre de Cristo mountain ranges.

In 1884, in the adobe house he built with his own hands, Felipe Cantu witnesses the birth of his daughter, Maria. Thirty-two years later, in the same house, in the same room she was born in, Maria gives birth to a son.

Strong and smart and tireless, Henry Salazar grows up to be a man of great resolve. Which is why when he meets a pretty, young fellow student named Emma Montoya at Capitol Commercial City College in Santa Fe, he doesn't let time, distance, World War II, or anything else get in the way of his pursuit. He *knows* she is the one.

During the war, Emma goes off to Washington to work as a secretary at the Pentagon; Henry goes off to Hawaii to serve his country. When his father suffers a stroke, Henry returns to the farm at Los Rincones, five miles from the town of Manassa. By now Emma is working in Albuquerque. One day, she comes home and a cousin tells her that a handsome guy has stopped by to say hello. It's been six years since Emma and Henry first met; he isn't about to wait much longer. They marry on Thanksgiving Day, 1948. Henry catches wild turkeys for their wedding dinner.

Henry and Emma live in the two back rooms of his parents' home. For creature comforts there are a wood stove, an outhouse, and a fifty-gallon barrel for water. Then the blessings begin to mount.

Leandro arrives first, in 1949. Fourteen months later, it is LeRoy's turn to enter the world. After him, in roughly two-year increments come John and Ken. Four years later Elaine is born. In 1961, Emma gets pregnant again. This time it is different. This time she becomes sick, very sick, toxemia sick. It isn't until after Margaret and Elliott and June are delivered that the doctor finds the cause of the toxemia: a fourth child died in her womb months earlier.

Emma recovers. The tiny triplets survive. But the medical bills are staggering, and Henry has no health insurance—what farmer does? The hospital agrees to a payment plan, but for a man of the land whose family has jumped from seven to ten in a single night, the crops and the small herd of cattle aren't enough. Henry finds a job at the Port of Entry in Fort Garland, the swing shift. He knocks off at midnight, drives back to the farm in his clunky 1950 Dodge pickup, and starts hand-irrigating the alfalfa by one thirty. He goes back to the truck, sets an alarm clock for an hour, falls asleep, wakes up, changes the water and goes back to sleep until it is time to get up and do it again. It is good he can get by on four hours of sleep a night. If he isn't irrigating, he's checking every two hours on the cows about to calve. Or doing something else. There is always something to do on the farm, always work to be started or finished before he returns to the house to shower, eat lunch, and grab the supper Emma has prepared for him. Then he heads back to Fort Garland for the next shift.

It is a backbreaking schedule, but, of course, Henry has help. There are many things that Emma and Henry teach their children—love of family, love of God, love of the land—but always looming large is the work ethic. Everybody has chores to do: feeding the animals, herding the cattle, working in the fields, picking potatoes, helping with irrigation, keeping the house clean.

Another thing the children are taught is that education is crucial. No matter how grinding his workload around the farm, Henry would rather die than take his children out of school to help him. Once, he wanted to earn his own college degree, but his father's stroke ended that dream. Nothing will end the one he has for his children. If they want to be farmers and ranchers like him, that is fine, that is good, but they will have their education first.

Henry has another dream. He dreams that at least one of his sons will be a priest. All but Elliott attend a seminary for at least a few years. Only Leandro comes close to being ordained, earning a masters degree in theology, taking his vows as a Franciscan brother. For the others, the path to God will not be made wearing a collar. As for the reason, well, as LeRoy would say many years later with an impish smile, "Maybe we all liked girls too much."

Not that there is much time for socializing outside the family. Between school and the farm, the Salazar children are kept busy. And when they aren't busy with studies or chores, they are busy at other things. They play together, fashioning toys out of sticks, riding horses, building rafts from old lumber, tossing a basketball into makeshift baskets, competing with each other to see which "team" can build the best storage cellar for potatoes. They grow up not just as siblings. They grow up as friends.

At night, the family gathers and the children do their homework in the glow of kerosene lamps—the only electricity in the house comes from a small gas generator and, since gas is expensive, that is used sparingly. Electricity will not come to the Salazar home on a regular basis until 1981, and neither will many of the comforts their peers take for granted. For all of their growing-up years, the Salazar children live without a telephone or television. When schoolmates talk about the latest episode of *Bonanza* or *Gunsmoke*, they may as well be talking about life on Mars.

But in the absence of TV, there are the *cuentos*, old stories read or told to the children by Maria, their grandmother, who lives with them. There are tales of the Salazar family, there is conversation, communication, there is the language of their heritage, Spanish, all flowing throughout the house. There are evening prayers and blessings bestowed by Emma. There is church every Sunday, and in between Sundays there is Henry's reminder about faith for all to see every time they leave the farm—the sign he has built and hung high over the entrance to their land, the one that reads *Vaya Con Dios*, Go With God. There is the slow intravenous drip of culture, of religion, of history, of the unbreakable connection to the land—all of it going into their veins, joining them to the past, preparing them for the future.

They all listen, they all learn, they all absorb what one of them would call the "intense love" of their parents. But no one listens or learns better than the second-born. And everybody knows why. "Henry Junior," that's what they call LeRoy, so alike in temperament are father and son. So driven. So impatient. So wedded to hard work and results. If LeRoy only has time for one sport in high school—track—well, at least he can run the 100, 200, 400, and the mile. At least he can set a school record in the mile, a record that will stand until Elliott comes along to break it.

Yes, there are high expectations—for all the children. A Salazar needed to accomplish things, be a winner. *If you run a race at school, come home and tell us about it if you finish first, otherwise, don't bother. Get good grades, get the best grades.* The bar that Henry sets for his children is high but never too high. Every one of them earns a scholarship, every one works his or her way through college and earns a degree. Four obtain masters; one opens her own business, one works with Cesar Chavez on the front lines of social justice, one becomes a U.S. congressman, one becomes a U.S. senator. But maybe no one *wants* to meet those expectations more than LeRoy. His sister Elaine thinks he is "the smartest one in the family." He is the one she calls "brilliant." The one his mother describes as "always very efficient." The one a friend calls "a very intense person, not easygoing, and definitely not funny."

The one who is second-born but fated to become the eldest in one terrible moment.

The ATV is pushing forty miles per hour because the man driving it doesn't know any other way but full speed ahead.

"My wife says I have two gears," shouts Salazar to the passenger behind him, "a very high gear, and a very low gear that's almost out of gear." In other words, when he isn't moving fast, he's probably sleeping through the night. Or taking one of his famous power naps. In the midst of the day's noise and light, he can go down for twenty minutes, a half hour, and bounce back refreshed. But being awake means being at work. Vacations are a chore for him, much harder than labor. And when you work, you must focus. "I like to get things done quickly and efficiently," he says above the thrum of the engine. "I can't stand to have people around me who move slowly."

Already, the ATV has zoomed past emerald fields of alfalfa and barley, past the large bins that store the varieties of potatoes that are the principal crop of Salazar Farm's 1,200-acre domain. For reasons having to do with financial arrangements and blood, the Native Son really can't break down which family member owns how much; everything seems to be tied together, especially now that he's essentially running John's operation while his younger brother toils away in Washington, D.C., representing Colorado's Third Congressional District. And if LeRoy is the force behind the herd of 150 Angus-Limousin cattle, you will find only his surname linked to Salazar Natural Beef because only the surname matters. "Ranching holds the family together," he says. Land and blood. Blood and land.

He points to the herd, and says with pride—but not vanity—"They look pretty healthy and contented, don't they?" It is a modest herd, he knows that. Larger than the twenty or so his father usually ran, but no great shakes in the firmament of Valley ranches. Still, if the animals are a much smaller part of the family business than the crops, they have their place with the Salazars. In the pasture, the mother cows and their babies are strolling, eating, staring back at LeRoy, who returns their gaze with affection. "We love the cattle," he says. "They're part of our culture, part of our heritage. We wouldn't ever think of parting with them."

The same can be said of Viking. Thirty-five years old, "the oldest horse in the Valley," Viking has a secure place with the family. No longer useful, a tumor in his eye and a languor to his step, he is still an icon, an elder to be respected and cared for. A creature to be honored and appreciated.

The ATV splashes over Sisneros Ditch so LeRoy can inspect the flood irrigation he is paying someone to oversee for him. His visitor asks an idle question about the longevity of the pasture being irrigated, only to find out there are no idle questions around the former teacher and agricultural consultant. *Let's see . . . cows will eat three percent of their body weight a day . . . depending on the size of the cow . . . times the number of cows . . . varies depending on conditions . . . whether the cow is lactating . . . hmm . . . two and a half tons of grass a day . . . that many pounds. . . .* Then, after mere seconds, he turns and announces, "That works out to about forty-five days of grazing we'll have on this pasture."

His mind seems to be most comfortable when it is working like that, humming along, confronting a problem, solving it, then taking on a new problem, a new challenge, as soon as it has dispatched the one at hand. Spend just an hour with him and you won't be surprised to hear him say, "I wish I had twelve different lifetimes so I could do twelve different things. Like what? Well, I've already been an engineer. . .teacher. . .ag consultant. . . ."

But this time he doesn't finish the thought because he's already dismounting the ATV and walking over to talk to the group of local teenagers—frequently some of the "rougher kids"—who work for him part-time. Not only do they have to work hard—and probably fast—they have to attend school and pass. No exceptions, no excuses. Salazar speaks with deep pride of one boy who used to skip classes before he came to work for El Rancho Salazar. How that boy now goes to school in the day, then works on the farm in the evening. How he'll take a load of hay that he's bought and drive it all the way to Santa Fe and make about 300 bucks selling it on his own. How he'll then drive back, get home to La Jara about three o'clock in the morning, and *still* make classes.

Maybe he sees something of himself in the boy, in his work ethic. In the ability to subsist on little sleep. He's matter of fact but proud when he tells his visitor about not needing more than that five hours of shuteye. More than that and he gets "antsy"; if he sleeps past four thirty in the morning it feels wasteful. But just as soon as he tells you that he only needs five hours, he feels the need to tell you that his father only needed four. Just like he seems to feel compelled to tell you "My dad had a way with cows; he had a sense of knowing what they were thinking, what would make them do things. He had a much better sense than I have."

He can talk about Henry for a long time, about the way "he had sayings for everything." *Dichos*—that's what they were called, aphorisms in English. Henry was full of them, LeRoy remembers. "*El que tiene, no pierde*—only those who don't have anything don't lose anything." He is smiling as he tells the visitor about how his father "was careful about what he'd buy; he didn't like to buy junk. He always said, '*Lo barato cuesta caro*—that which you buy cheap ends up costing you a lot more in the end.' " He can talk about how both he and his dad can be called "hyper, we're both workaholics, very demanding of ourselves, very demanding of everybody around us. I always tell my young folks, 'One of the reasons I'm successful in life is I surround myself with good people.' So I don't want anybody around me who isn't going to be willing to work real hard." Sometimes, he knows, he can be too demanding, too hard on those around him. Sometimes there is too much Henry Junior for people to take. But he tries to ease up on the throttle; he has learned not to erupt as easily as his father did. Even if Henry Salazar was quick to calm down, his temper could feel like a lash.

When the Native Son isn't talking about his father, he's happy to talk about his mother. How, "My mom, she has the most positive attitude of anyone I know. She has this generosity about her, this spirit, this optimism. When I think about living the Gospel, I really think of my mother. She's the only person I've ever been around for an extended period of time who never says anything bad about anyone. She's the least judgmental person I've ever known in my life."

And when he's not talking about his mother, he'll talk about his siblings. About how "Elaine and I are a lot alike." About how Ken, the U.S. senator, and John, the U.S. congressman, "are what some people call 'salt of the earth.' What you see is what you get with them. Good rural values. Family, faith, and community. That's what we were all raised with."

When he's not talking about his siblings, he'll talk about Michelle, his wife, about what a "great teacher" she was in high school and college, about how proud he is that she's finishing up her Ph.D. dissertation in Hispanic linguistics. Or he'll talk about his three grown children—Rosie, 31, Gabe, 29, Lucas, 27—and how proud he is of them. How Rosie works for a radio station but is really a "dancer, an actress." How Gabe is a software engineer with a son. How Lucas is a teacher but also plays in the 58 Cent Band and "you should look him up on the Internet, on YouTube."

And when he's not talking about his father, his mother, his siblings, his wife, or his children, he still will try to avoid talking about himself because that makes him uncomfortable. His story is simple and not so interesting. At least not to him. No, what is much more interesting is talking about Andrew, his three-year-old grandson, "the great joy of our lives." About how the boy "loves to be with me at the farm." For Salazar, there is nothing more wonderful than this mutual love he and his grandchild have. For to love the farm is to love the land. And to love the land is just *natural*.

He has stopped his truck and gotten out, no small concession from the man who hates to stop working. But he has been asked about the land and he needs to make sure his visitor understands.

"I love nothing more than to dig down into good, rich soil and smell it and feel it with my hands," he says in a voice that is practically rhapsodizing; the voice of a man talking about his lover, the voice of a priest talking about his deity. "When I'm away from the farm and I have to go to the city for a while, I feel like something is just missing in me."

For a moment he is quiet.

"This is the way man was meant to live," he says. "Being close to the land is important for all of humanity. I wish more people in the city had the opportunity to see this." He is looking out at the immense Valley, where the wind is starting to rage again, scouring the canopy of sky that is stretched taut between the faraway mountains. "Living on the land is important, passing it on is important. Living on land that's been passed from generation to generation gives you a sense of history, a sense of groundedness. This ranch, it's part of the fabric of our lives." He looks at his visitor, eyes intense, as if he is hoping to will him into understanding what he is about to say next. "Because the land is what made us who we are."

Since he has already stopped the truck and gotten out, it seems inevitable that LeRoy Salazar will quickly find some new task that is close at hand, something to occupy himself with while he talks. He steps around to the rear of the vehicle and plucks a shovel from the bed. Then he walks maybe ten yards and begins attacking a thicket of weeds that are growing around the electric pump that drives water into a center pivot irrigation system. That done, he gets in the truck and drives to the center pivot. He climbs out and approaches the large skeletal structure that is really what makes it possible for the Salazars to farm in the cotton-dry air of the Valley, where the average rainfall is barely six inches. He stands at the computer panel and as he programs it, attempts to explain the concept behind it, how it works, what it does, the scientific harmony of it all. For the moment, he is a teacher, the kind of solid, passionate professor he might have become thirty years ago if he had gone for his Ph.D., like two universities wanted him to. But he hadn't wanted that. He hadn't wanted to become rooted in the brick and ivy of academia. He wanted to be outside, in motion, working on the land with people who weren't slow.

He wanted to come home.

It is not easy for a man to compress three decades of his life into less than an hour. Unless, that is, the man is LeRoy Salazar and he's got work outside, waiting to be done.

So, sensing the clock is ticking on the sit-down portion of their meeting, his visitor quickly figures out how to decipher Salazar's verbal shorthand. He listens without interrupting and learns how Salazar used a scholarship, loans, and odd jobs to work his way through Colorado State University, where he earned a degree in agricultural engineering because "I loved farming and ranching and was good in math and science." He learns how LeRoy then spent two years in the military, in the U.S. Army Corps of Engineers, and returned to CSU for his masters in—what else?—ag engineering. Somewhere between degrees, he met and married Michelle Lankford, even though, he says, they only had one "formal date." For the record, it was a double date. For the record—Michelle will tell you—"He never proposed. It was pretty much understood. The only question he asked, and it was over the phone, was 'So when are we getting married?' Something like that." Just like Henry knew when he first saw Emma, LeRoy *knew* when he first met Michelle.

He spent several years as an agricultural consultant, teaching irrigation and soil and water conservation in places as far away as South America and India. Even though the consulting took him out of the country for months at a time, he always made sure to keep April and October—farmers' months—free to help his father and brothers with the planting and the harvesting.

By 1980, he had come to a crossroads. CSU and Utah State University were both trying to woo him for their Ph.D. programs. Down that path, the path of academia, lay a life of teaching and research. A safe life, a good life.

But not the life he wanted.

For one, he was a "hands-on person," someone who wanted to be applying agricultural technology in the real world, not the insular world of a university. For another — and, make no mistake, this was the *driving* reason — he wanted to come back to the Valley. He wanted his children — Rosie and Gabe had arrived, and Lucas would be coming along soon — to grow up where he had grown up. Wanted them to hear the songs of the animals and feel the soil in their hands. Wanted them to learn what had made the Salazars who they were.

But by now, Leandro had returned to the Salazar fold, finished with his organizing work for Cesar Chavez. And with Henry and Emma and John and his family already settled on the land, the farm could not sustain LeRoy and his brood. No matter. He founded Agro Engineering, an agricultural and agronomic consulting firm based near Alamosa. He worked hard. His business prospered. His children were near their grandparents, near the land. A decade rolled by. Another was in progress. Everything was going well.

Then, on April 22, 1992, Emma's seventieth birthday, it happened. So quick, so black. A nightmare in the morning light.

Leandro is in the field, driving the tractor, preparing the land. He gets out to fix one of the large implements the tractor is pulling. He doesn't turn the tractor off — you never do — just lets it idle. He bends down. The wind is blowing. It catches a flap of Leandro's jacket. He doesn't notice the tractor has started slowly rolling towards him. The jacket snags on a large hydraulic shaft. Leandro can't get free. He is pulled against the shaft. All the air is squeezed out of him. Then all the life.

They buried him in Los Cerritos Cemetery. On the grassy slope near his grandfather and great-grandfather. The family buckled, but held together. Emma was the rock. Henry was not. God gives and God takes away. But this?

The wind blew the years across the Valley. By 2000, Emma and Henry were in fading health; all the siblings could see that. Both had heart conditions, and their father's Alzheimer's was starting to worsen. For LeRoy, the decision to phase himself out of Agro Engineering and move back to the ranch full-time was not hard. It was not hard to come back and help his parents, to see the calves being born and raise life out of the ground, to assume the responsibilities of the oldest child. Blood and land. Land and blood.

Christmas 2001 was approaching. LeRoy and his father were done with another good day of sharing work. Despite his heart, Henry was still strong, still capable of lifting bales of hay, still capable of carrying firewood and feeding the animals. That, in fact, is just what he had finished doing when he went into the house to shower. In the shower he had a heart attack.

San Luis Hills

When the ambulance came, Henry struggled. He did not want to go to a hospital. He wanted to die in his home. He continued struggling at the airport as they put him on a plane. He wanted his last breath to be taken from Valley air. But it was not. A day and a half later, he died in a hospital in Denver.

They buried him in Los Cerritos Cemetery. Not far from his oldest son, under a large headstone. Next to Henry's name on the stone was Emma's name; she wanted it to be ready for her when the time came. On the back of the headstone, etched into the granite, were the names of all eight of their children. Because with some family histories, death is only a footnote, not the final chapter.

It is afternoon-bright around the table in the house where Emma Salazar has lived for nearly sixty years. She is eighty-six and her health is not good. For ten years, they have been telling her that. For ten years, she has been coexisting with a bad heart. For five years, she has been recovering from a cerebral aneurysm. And yet her eyes remain bright and hopeful and the shy smile she shows the visitor looks like it belongs on the face of a pretty, young woman — the same pretty, young woman you see in the framed photograph that sits next to so many other photographs of family members both dead and alive. She is still the luminous candle of the Salazars, the one whose counsel is sought, whose kiss is a benediction. The one who still offers a blessing for any family member who is about to leave the house after a visit. The one each of the seven siblings calls everyday. The one who always — always — has one of her children present in her house at night when she sleeps.

Right now, she is talking to her oldest living son and telling him, "*Mi hijito* — My dear son — you do work too hard."

"But I'm happier than I've ever been, mom," LeRoy tells her. "Work is my passion."

She smiles. She looks at her visitor and says, "I am very lucky to have my children."

"No, we're the lucky ones, mom," her son tells her.

Soon he has to go outside and return to work. He will be back tonight, of course, to check in on her. To share evening prayers. To be there. To return to her house because all the roads he cares about lead here.

He hopes someday his children will return to this place. He thinks they will, because he knows that his children, and the children of his brothers and sisters, "Care enough about the legacy, care enough about keeping the family together, care enough about the values that they grew up with, that they will do their best to keep this land of ours together. Yes, I do think that some of them will come back someday."

It would be a shame if a sixth generation of Salazars didn't come back to the land, didn't return to keep it productive and beautiful, but one day this will no longer concern the Native Son. One day he knows that, just like his great-grandfather and grandfather and father and brother, he, too, will be dust. Another name etched on a headstone. Another grave lying under the green grass that grows in the Valley that is like no other. The Valley where land and blood are forever linked and a man can hold paradise in his own two hands.

[On Jan. 20, 2009, Ken Salazar was sworn in as the U.S. Secretary of the Interior.]

Ranch headquarters

HILL RANCHES

- *Saguache County*
- *Owners: Gary Hill family*
- *First ranched in 1874*

Hill Ranches' Middle Creek Ranch lies along Saguache Creek six miles west of the town of Saguache. It is defined by its large hay meadows and the numerous meanders of the creek, as well as the Bureau of Land Management administered volcanic foothills and mesas that confine the valley. Several tributaries, including Middle, Jacks, and Ford Creeks, drain through the ranch into Saguache Creek. Six miles to the north, the Continental Divide runs along the crest of the Cochetopa Hills, separating the ranch's Rio Grande headwaters from those north of the divide that ultimately drain into the Colorado River.

Middle Creek drainage

ROCKY MOUNTAINS

FLYING X RANCH

- *Saguache County*
- *Owners: Anne Curtis Nielsen and Ed Nielsen*
- *First ranched in 1884*
- *Conservation easement held by Colorado Cattlemen's Agricultural Land Trust*

Flying X Ranch, thirteen miles west of the town of Saguache, lies along one of the more scenic river meanders anywhere in Colorado. Saguache Creek drains from the La Garita Mountains south of the ranch and bisects the property in the valley bottom on its way east to a disappearance into the San Luis Valley's underground aquifer. The creek meanders, vast hay meadows, and the volcanic rock flows that contain the valley define the scenic qualities of the Flying X. An historic stage coach route over Cochetopa Pass that connected the San Luis Valley with the town of Gunnison passed through the ranch. Part of a 19th-century log stage stop still stands on the ranch. Flying X was first ranched in 1884 and was purchased by the Curtis family in 1951. Anne Curtis Nielsen is the fifth generation of a family that began ranching in the San Luis Valley in the 1860s. Ed Nielsen's family ranched the Meeker and Rangely areas in Colorado.

Saguache Creek

Saguache Creek meadows

Saguache Creek and ranch headquarters

RIVERGATE RANCH

- *Gunnison County*
- *Owners: Dave Gorsuch family*
- *First ranched in the 1880s*
- *Conservation easement held by Colorado Cattlemen's Agricultural Land Trust*

Rivergate Ranch lies fifteen miles north of the town of Lake City along the banks of the Lake Fork of the Gunnison River. The river bisects the ranch and the hay meadows that define it. Cottonwood trees line its banks, making intense color in spring when they are first leafing and in autumn when leaf colors turn yellow and orange. Rivers in the West cannot be much more scenic than this one. The Uncompahgre Wilderness can be seen immediately to the southwest, and the Powderhorn Wilderness, which serves as summer grazing territory for the ranch cattle, to the southeast. Ute Indians used the land that became the ranch as a summering camp. The ranch later served as a fishing retreat for many years. The Gorsuch family acquired it in 1994 and today run it as a beef cattle operation.

Twilight above the Lake Fork of the Gunnison River

Ranch overview looking south

Evening primrose along the Lake Fork

RIO OXBOW RANCH

- *Mineral County*
- *Owners: Alan and Patsy Lisenby*
- *First ranched in 1897*
- *Conservation easement held by Rio Grande Headwaters Land Trust*

Rio Oxbow Ranch lies half-way between the towns of Lake City and Creede in Antelope Park along the Silver Thread Scenic Byway in the headwaters of the Rio Grande. The river bisects the ranch and its broad grazing meadows that fill the valley. Volcanic peaks—like Bristol Head to the north—and the foothills of Colorado's largest and most remote wilderness area, the Weminuche, contain the valley. Significant wetlands are home to waterfowl. Formerly called Wright's Lower Ranch, it was owned by the same family for one hundred years until the Lisenby family purchased it in 1997.

Ranch relic!

Geese flying

Rio Grande

WILSON RIO GRANDE RANCH

- *Rio Grande County*
- *Owners: Virginia and Roland Wilson*
- *First ranched in 1874*
- *Conservation easement held by Colorado Cattlemen's Agricultural Land Trust*

The Wilson ranch lies three miles west of the town of Del Norte along the banks of the lower Rio Grande on the edge of the San Luis Valley. The river is relatively broad at this point, and it nurtures massive century-old cottonwood trees in its flood plain. It is a sight to see them turn yellow and orange in the fall: they make a conspicuous wall of color for many miles along U.S. Highway 160 to the south. The ranch contains large grazing pastures south of the river and foothills of volcanic rock to its north.

The land was ranched as early as 1874—the Wilson family possesses the original land grant signed by President Ulysses S. Grant—and was acquired by one Emil Von Bernuth in 1893. Not only was it ranched for cattle, fish ponds were constructed in the 19th century to raise fish for consumption. The Wilson family began buying parts of today's ranch in 1979. During the ensuing years they improved the ranch, including restoring the ponds to productivity, refurbishing a 19th-century barn, and importing from the San Luis Valley an historic cabin built in the 1860s that was slated for destruction.

Historic cabin, built 1860s

Pronghorn antelope

Rio Grande

Evening twilight

WEMINUCHE VALLEY AND NOTCH RANCHES

- *Hinsdale County*
- *Owners: Robert D. Lindner family*
- *First ranched around 1902*
- *Conservation easement held by Colorado Cattlemen's Agricultural Land Trust*

Both of these ranches lie within the headwaters of the Piedra River, twenty miles north of the town of Pagosa Springs in pristine mountain parks. More importantly, they are tucked up against the south side of Colorado's largest wilderness, Weminuche, and within a stone's throw of a remote section of the Continental Divide. Various tributaries of the Piedra, including Weminuche Creek and the Middle and East Forks of the Piedra, run through the ranches year-round. Grass meadows interspersed with cottonwood trees define Weminuche Valley Ranch, while Notch Ranch boasts as beautiful a ponderosa pine forest as seen anywhere in the West. Craggy volcanic ridges serve as the backdrop for both ranches, and 13,000-foot peaks along the Continental Divide can be seen farther in the background. The ranches were homesteaded early in the 19th century by various families, including the Shaw, Davis, and Clayton lineages. Robert D. Lindner acquired Weminuche Valley Ranch in 1971 and Notch Ranch in 1983.

Blooming corn lily

Ponderosa pines

EL RANCHO PINOSO

- *Archuleta County*
- *Owners: Robert D. Lindner family*
- *First ranched in the late 1880s*
- *Conservation easement held by Colorado Cattlemen's Agricultural Land Trust*

El Rancho Pinoso, as the name suggests, contains extraordinary old growth ponderosa pine trees. It lies fifteen miles to the east of the town of Pagosa Springs. The Rio Blanco drains from the South San Juan Wilderness immediately to the east and bisects the ranch. Just like its sister ranches Weminuche Valley and Notch, also owned and operated by the Lindner family, El Rancho Pinoso is just below the Continental Divide, and volcanic ridges and 13,000-foot peaks create an extraordinary background. Its vast meadows create one of the largest private mountain parks in Colorado. Cottonwood trees along Rio Blanco complement the stately ponderosa pines in the higher parts of the ranch. Robert D. Lindner acquired the ranch in 1986.

Ponderosa pines

HARTONG RANCH

- *Archuleta County*
- *Owners: Hartong family*
- *First ranched in 1888*
- *Conservation easement held by Colorado Cattlemen's Agricultural Land Trust*

Hartong Ranch lies twenty miles southeast of the town of Pagosa Springs below the Chalk Mountains of the South San Juan Wilderness. As one ascends various four-wheeler trails, its hay meadows give way to dense forests of scrub oak, ponderosa pine, spruce, fir, and at higher elevations, beautiful groves of aspen interspersed with lush meadows. Wildflowers seem to bloom everywhere in June and July, and the aspens make for a colorful spring and fall. The ranch was homesteaded by the Cornish, Paximan, and McBroom families in the 19th century. Ben and Mimi Gardner purchased it in 1936. Their daughter Drue and her husband, Chuck Hartong, took over ranch management in 1954, and today her son Ron oversees it.

She doesn't hear too well, but that hardly matters because the loud mechanical gargle of the four-wheeler is swallowing up anything that isn't a shout. And she doesn't always remember things as quick as she'd like to, but right now that doesn't matter much either because as long as she can recall the sites of all those dips and ruts and jagged scars in the road—and can see them with those clear blue eyes—well, everything will be okay.

At least you hope so.

You hope so because here you are 8,500 feet above sea level, hugging a mountainside, sitting on the back of that four-wheeler, bouncing up and down on your ass, holding on with numbing fingers as an eighty-five-year-old woman with a sugary Texas drawl and snow-white hair that's shining in the sun chauffeurs you across some of the prettiest country nature has ever come up with.

"You okay back there? You haven't fallen off yet, have you?" she shouts, slapping away a low-hanging branch, adroitly navigating down and up a deep crease in the washboard road that her husband pretty much built by himself fifty years ago. "I'd really like to take you down to the river. That's a way to get you up to the Mac-Broom place. It's really pretty up there. Do you wanna try?"

Well, of course *you* want to try because you know that *she* wants to try. And even the most unrepentant city-slicker son of a tenderfoot can figure out that when you're riding with a feisty, zesty, pull-no-punches great-grandmother who was weaned on sandstorms and suffers fools poorly the last thing you want to do is disappoint her and get on her bad side. Besides, you also got to figure that if cancer, a tricky heart, a missing toe, an arthritic hip and that fancy-pants doctor who said she only had one year left to live two years ago haven't stopped her, neither is one more four-wheel plunge to the river.

So—*whooey* and *yeehaw*—you shout out, "Sure," and down to the Little Navajo River you go, ducking under branches, casting your fate to the curling wind and two small hands, comforted by the lurking suspicion that, whether you're talking Chromo, Pagosa Springs, Archuleta County, or, shoot, probably the whole cotton-pickin' state of Colorado, there isn't anybody around with more gumption or steel than Emma Drucilla Gardner Hartong.

And maybe there never will be.

"Huh, if it'd been a snake, it would've bitten me," says Drue Hartong, locating the creamer that has been staring her in the face for ten seconds. She puts it on the table, fills it with milk and encourages you to put some in your coffee, even as she's coaxing you to put some cold chicken on your plate for lunch.

"I had ham, but Ron ate it all," she says, shaking her head disapprovingly at the notion that her son finished the lunchmeat. "Remind me to cuss him when he comes home," she adds, a twinkle—yes, it is a twinkle—in her eye and a smile on her lips. "You'll just have to make do with this chicken. Say, is that all you're going to eat?"

Uh, of course not.

Aside from taking an adequate helping of chicken, the first thing you want to do to avoid getting on Drue Hartong's bad side—"If that happens, forget it," says one lifelong friend—is never call her anything but Drue. Drucilla might be okay, but she *hates* the name Emma, even though it was her grandmother's name. The second thing you want to do is make sure you don't mention mules to her. "I can't stand to even look at a mule," she says, never forgetting that a mule once kicked her dog "out of pure orneriness," and that her father once told her, "Never trust a mule. They'll live 100 years to kick you once."

The third thing to remember is don't bring her any more cats. She's got nine now, and "I didn't ask for a single one of 'em." People just keep bringing them by because they know that a soft touch like Drue will surely feed them.

The fourth thing to keep in mind if you want to stay on Drue's bright side is you better be careful about recommending books. She loves to read, but "You sure have to be careful what kind of book you pick up. A lot of books, I read about the first chapter and then it gets thrown away. Sex, sex, sex—that's all they wanna talk about."

Finally, be prepared to patiently repeat questions to her at a loud volume and don't always expect a quick answer because "My hearing's pretty much shot and my memory's gone bye-bye." She says this without a hint of self-consciousness or self-pity, just as matter-of-fact as you can. Which is pretty much the way she is about everything.

Take her cancer. It's melanoma for sure. They found it in her foot and had to remove her big toe. Found more in her lungs where "it's growing like a weed." Radiation and chemotherapy don't work against melanoma, which makes it pretty scary to most people who aren't Drue Hartong. She asked the doctor how long she had left and all he did was stutter and mutter something about a year. That was two years ago and she's still here to say, "I'm not gonna cry about it. It's not gonna bother me one way or the other. I've done everything I really wanted to do anyway."

Truth be told, she sounds way more upset about her arthritic hip. It's just worn down to the bone and can get to hurting so bad "There are times I'd like to take a knife and cut the damn thing off. But then I'd be in a fix, wouldn't I?" Yeah, that bad hip put the "squash on pretty much everything," but mostly what it did was make it impossible for Drue to sling her leg over the saddle. And, oh, did she love to ride horses. Why, as a teenager, she once rode all the way from Chromo to Pagosa Springs—that's twenty-two miles, sir—and didn't think anything of it. But, of course, she'd been riding practically her whole life. Or wanting to. She can still remember all those long-ago Christmas mornings, when she'd spring out of bed and look out the window, hoping her present was a horse, thrilled down to her toes when one day—and it wasn't even Christmas—Sugar was waiting outside for her. She was riding bareback the very day she first laid eyes on the ranch, too. That was back in 1936, when the dust storms in Texas had finally convinced Mimi and Dad there had to be a better place to live.

Okay now, whoa. Let's stop here for a minute and do a little bit of tending to this story. No, not tending like in spraying the Canada thistle up along the Paximan Place, or mending busted fences higher up in McBroom country. Instead, let's try to make a little bit of order while we assemble the mosaic that is Drue Hartong's life. Of course, since she's absolutely plainspoken—and you better be, too, when you're around her—maybe mosaic is too highfalutin a word. Instead, let's just call it a jigsaw puzzle and start putting a few pieces together.

Plainview, Texas, was dry and brown but that wasn't the worst of it. No, the dust storms were. You'd go to bed at night, close the windows and shut the door, and, even then, when you woke up in the morning and lifted your head off the pillow you'd see a white silhouette surrounded by all this fine sand that'd been blown in through the cracks. Yessir, Plainview seemed like a fine place to get away from, at least from a teenaged girl's point of view. So when her mother and father said that they were all taking a vacation to Colorado that summer of nineteen-and-thirty-six, Drue was pretty excited.

They drove up through New Mexico and got into Colorado near Chromo. It was getting dark, so they asked some folks if they could just camp on their property for the night. They found a spot right near the Little Navajo River, where the water rushing past in the darkness sure sounded a lot better to Drue than those shrieking winds back in Plainview. Next morning, Mimi—that was Drue's mom—asked the nice people if they knew of any property for sale. Well, now that you mention it. . . .

They rode up on borrowed horses, shy one saddle so Drue rode bareback. They rode up through the gamble oak, chokecherries, and wild purple irises, past the beaver ponds and the grass so green it almost hurt your eyes, up toward that V-shaped mountain and, lordy, they were just "overwhelmed" by the beauty. Compared to dingy, dusty, windblown Texas, finding the ranch was like finding pieces of rainbow all over the ground, just waiting for you to stuff into your pocket.

After Mimi saw the ranch, Plainview was pretty much history. Ben—that was Drue's daddy—sold the house, his garage, and whatever else he could, and pretty soon the land that had been worked by the Cornishes, the Paximans, and the McBrooms—all homesteaders whose sweat had tamed some of the 3,000-acre property—belonged to the Gardners. Ben liked sheep and he soon had himself a herd. But he kept a few cows, too. He'd milk them, separate the cream out, pour it into a five-gallon can, tie that can to his horse and ride the two miles to Chromo so he could sell it. He could've driven the family car, but he preferred riding.

Meanwhile, Mimi had taken one look at the schools in Pagosa Springs and said uh-uh—no way was her daughter going to be educated there. Mimi was a good secretary and quickly got a job working for a lawyer in Durango, about sixty miles away. Mimi and Drue got an apartment and lived there during the week, Mimi working for the lawyer, Drue going to school. In the summers, Drue would stay at the ranch, managing not to kill any of the hands with her cooking, and riding as much as she could, practically living in a saddle, growing to love the land. The Chalk Mountains, the Little Navajo, the velvety green pastures—this was where she was meant to be. There was no plumbing and no central heating in the house, but all you had to do was step outside and look around to be reminded that staying close to the wood stove in winter and fetching water from the spring wasn't all that bad a tradeoff.

Drue went off to college, first in New Mexico, then in Texas. Then she headed to California to work for the airlines as a stewardess. That was where she met Chuck Hartong. That was where she married him, too. And, since we're on the subject, that was where she bore him three children—Chuck, Jr., and then the twins, Ron and Karen. Chuck and Drue and the kids would visit the ranch as often as they could. Drue could tell that Chuck, an auto mechanic by trade and a city boy by birth, saw the same pieces of rainbow that she did. That's why they both knew there was only one way to answer when Mimi gave her ultimatum.

Ben had been doing poorly for years. So poorly, the sheep had already been sold by the time he had his heart attack and died. It didn't take Mimi all that long to realize she couldn't handle the ranch by herself. If Drue didn't want to move back, well, then Mimi was going to have to sell it. *Sell it?* Why, even when she was living in California, working, getting married, having babies, Drue always knew the ranch was her real home. *Sell it?*

In 1954, the Hartongs came back for good and got to work. Maybe Chuck was no rancher, but he sure picked things up quick. Learned to milk cows and, after he and Drue and Mimi decided to grow their own herd, became a pretty good cattleman. Learned about irrigation, too. Got himself a bulldozer and began sculpting roads up to the Cornish Place, the Paximan Place, the McBroom Place. Took a long time, but eventually you could get up there on something besides a horse.

But it wasn't just Chuck working the ranch solo. Drue sure knew what it meant to put in a full day by herself. And, as the kids got older, they were on a first-name basis with hard work, too. They learned how to take care of the cattle. They learned how to build fences. They learned to spray the Canada thistle that was always ready to take over a pasture if you didn't stay on top of it. Oh, they weren't perfect—Karen and Ron, in particular, were a handful for their mother. Lordy, what those two couldn't think of to get into. And most of the time it was Karen doing the instigating. Like that one winter she told Ron the ice over the Little Navajo was thick enough to stand on, only of course it wasn't and they both went through and got about as frozen-soaked as you can. "Well, she's got red hair, does that explain it?" is what Drue would tell people about her rambunctious daughter. Taking a yardstick to the twins every now and then helped some. But, trouble making aside, when it came to chores and hard work, none of the kids needed much reminding. They learned their work ethic from a mom and dad who never slacked off.

Life went on. The kids went to college. Chuck, Jr., got a good job in Texas and moved away. Eventually, he'd have kids and grandkids of his own. Karen got married and moved away, too. Ron mostly stayed to work the ranch. And even if he preferred tinkering with machines to riding horses—he'd always joke that he preferred his horsepower under the hood of a truck instead of under a saddle—he kept things humming over the years, even as changes inevitably came. Not that he did everything by himself.

Drue continued to pull her weight, too. Maybe she couldn't wield the calf pullers—"fetal extractors" was the fancy term—and help Ron deliver a calf that had gotten turned the wrong way in its mother. Maybe she never learned how to perform artificial respiration on a newborn whose lungs were full of fluid like Ron did. But she could sure roll up her sleeves and get her hands plenty dirty. She could help feed the animals. She could do some irrigating. She could make those Canada thistle wish they'd never been propagated. She could help build and repair fence, too. Something needed to be done, she'd find a way to make sure that it was. Sure, "when you're a woman rancher you can't move a big rock like a man, but you can figure out a way to do something you gotta do."

One thing she couldn't do was keep time from claiming who it wanted. The cancer finally ate up Mimi in 1983. Two decades later, after a morning of riding around on his four-wheeler and maybe getting a little irrigating done, Chuck, Sr., had a massive heart attack and died. By then, Drue had already slowed down, what with the stints they put in her coronary arteries in 1997 and the arthritic hip that was giving out.

Still, even as she kept adjusting to each new tap on the shoulder from old age, something wasn't setting well. At first, she ignored it. She knew part of the rancher's creed is "if it doesn't hurt real bad, you don't go to a doctor." Finally, when it did start hurting real bad, she went to the doctor. That's when they found the cancer.

But she wasn't going to cry about it.

"Would you look at that? Doggone it—somebody built a house up on that far hill over there, see?"

The new home, sticking up all peacock-proud in plain sight on a hill that, truth be told, isn't all that close, is enough to bring out the vinegar in Drue as she sits on the idling four-wheeler and vents her frustration. "Why can't they build it down by the river where you can't see it? Why do they have to go and spoil the beautiful land?"

By the time she's brought the vehicle back down and put it away in the shed and returned to her kitchen table, she's sort of calmed down. At least until some dunderhead asks her how she feels toward ranch folk who sell their land off to developers.

"Dirty, cotton-pickin' so-and-so's—I'd like to shoot 'em!" she says, which are pretty strong words for a woman who never liked to hunt and couldn't even stand to help out much at brandings because "you know it hurts. The little calves would bawl." But, oh, the thought of so much beauty being crowded by people and their fancy new homes, well, "Jiminy, it's enough to make you want to cry. You see all these houses and it makes you wonder if those people are gonna take care of the land like they should be."

"It's sure not the secluded solitude we're used to," says Ron, who has come in from hauling logs. He has a booming voice and the easiest laugh of anybody in Archuleta County. Doesn't take the slightest offense to the peevish tone in his mother's voice when she demands, "What'd you do with my four-wheeler? I had to take Karen's. What'd you do with mine?"

She's already in a somewhat contrary mood because he ate the ham. And on top of that, she's almost for sure vexed because she "just went blank" when trying to remember both the name of the Chalk Mountains and what company Chuck, Jr., works for in Dallas. But more than likely what's really got her dander up is that the Little Navajo was running too fast and full to cross, so she wasn't able to take her visitor up to the McBroom Place, which is "the prettiest place there is."

Ron doesn't know this, not that it would matter much if he did. He can read her pretty good most of the time. He looks her in the eye, raises his voice a notch louder to make sure she can hear, and says, "It's in the shop, mom. I was charging the battery up on it. I never have gotten around to getting it out."

His easygoing nature is probably a good thing, particularly on the days when the workload of the ranch gets to be too much, on those days "when everything goes wrong and even a hammer doesn't work." Today isn't that bad, but it sure hasn't been a picnic. See, he's been hauling logs for a shed he's fixing to build. But that means he didn't get to the irrigating that he needed to. Sure, he might do that tomorrow, but tomorrow's Saturday, and Karen, as she does just about every weekend, will be coming out to help around the ranch, and they're planning to fix fences, which means he might not get to irrigating then either. Which means he'll have to do it Sunday—or not, if something else comes up. At least he doesn't have to worry about the cattle anymore, even though that's what you could call a mixed blessing.

"I sure do miss the cows," says Drue sadly, propping up her head in her hand as she rests an elbow on the table. "They sure kept you busy. You're always doing something for them. But it was too much for Ron to do by himself. He was really dragging; he looked terrible."

The recent sale of their one hundred head of cattle is a source of sadness to both Ron and Drue, but especially to Karen, who cried when it happened. Ever since she'd moved back to the area in 1994 and began working for the U.S. Department of Agriculture in Durango, she'd been making the eighty-mile one-way commute from Ignacio to the ranch just about every weekend. Heck, she'd even take two weeks off from work every year so she could help out during calving season. Even after her divorce, when she became a single woman again, the ranch stayed her focus. The only vacation she took in fourteen years was to Jackson Hole, a pretty enough place, for sure. But after one week of Jackson Hole's beauty, she was antsy to get back and help Drue and Ron.

Still, Karen knew that coming in for the weekends and calving season wasn't going to cut it when it came to helping Ron with the cattle. So, this past spring, she reluctantly went along with the idea to sell the herd. Learned to deal with the emptiness that seems to hang around the ranch now that the cattle are gone. Good thing she and Ron managed to lease out some of their pasture to another rancher. That'll help until the day finally comes that she can start raising Hartong cattle again, although who knows when that will be? Not until she retires from her government job, that's for sure.

But whenever her retirement comes, at least she knows the land will be waiting for her. See, it isn't just Drue and Ron who want to keep the ranch as a ranch. Heck, Drue figures that Chuck, Jr., would give his eyeteeth to get back here and be working the ranch. As for Karen, well, it's practically a dead heat between her and her mother as to which one is the most gung-ho about protecting their property. Go ahead and ask her how she feels about the conservation easement that the family put on the land to put the kibosh on any future development, and she'll tell you she was behind it "one hundred percent! None of us want to see houses on this place. Ever." She'll tell you about all those years when you could be riding a horse down from the pastures and look out south toward the Big Navajo River and not see one single rooftop. She'll tell you that now you can see several and how there's probably going to be more than that if things keep up. After all, didn't folks like the Crowleys, the Bigbees and the Eaklors—good ranching families, all of them—sell? And then didn't houses start to spring up like Canada thistle? Well, that's not going to happen to the Hartongs. Thanks to that easement, says Ron, "This land is protected until eternity."

"And I'll die happy," snorts Drue.

Doggone right she will. And her spirit will stay even happier if Ron and Karen and Chuck, Jr., do like she told them and spread her ashes across the land. Preferably in the upper country, along the McBroom Place, but anywhere out on the ranch is fine. Because she wouldn't want to wind up anywhere near a city "for love or money." And while they're at it, they can spread their father's ashes along with hers. Right now, those ashes are "in a can. And you better believe I don't want to be shut up in that thing."

Saturday morning has arrived and so has Karen. She moves quickly and speaks quicker, probably because she's in a hurry to be outside and get some work done.

"I guess the most important thing mom and dad taught us is a work ethic," she says. She pauses. Wants to say it another way. "They taught us to, well, to work for what is yours, is the best way I can put it."

She looks over at her mother.

"I'm not hearing anything you're saying," says Drue, just as matter-of-fact as you please.

In a louder voice, Karen tells her mother what she just said. Drue smiles, but only halfway. She feels bad because she hasn't been able to help much for too darn long, if you ask her. Then again, if you do ask her, she's not exactly sure for how darn long.

"When's the last time I was up spraying with you?" she asks her daughter.

"Well, prior to o-six, I'd say you were up there ninety percent of the days in the summertime, Mom. Then you had the cancer diagnosed and that really slowed you up. But you helped a little bit last year," says Karen cheerfully.

"I sure do miss not being able to get out like I used to," says Drue, shaking her head, more vexed than sad. "You better believe it bothers me not getting out to help."

At least she can still cook for Ron; she enjoys that. She also enjoys the fact that "Ron'll eat anything."

There's that twinkle again.

"Yeah, that's one thing about ranching," booms Ron, as he walks into the kitchen, laughing and talking at the same time, "When you're working you're hungry. You don't have a finicky palate, that's for sure."

Yeah, but you better believe that Drue wishes she could do more than just cook. She misses not being able to help some with the calving or the cows, even though it's been years since she was agile enough to not get in their way so she wouldn't get stomped. She misses not being able to get up on her horse and ride to the McBroom country and maybe help break up cow pies. She misses spraying the dang weeds. She misses doing *something*.

Not that she's given up on the idea.

"I'm feeling a lot better this year than I did last," she says. "I think I'll be able to help more."

And she might. Sure, this year has been a little tougher. All that snow just flattened a good five miles or more of fence. And then there's the way the winter just lingered and lingered so that it felt "like the cold just didn't want to turn loose. Did you know it was zero in April?" Even yesterday, "The wind blew so bad I thought I'd go crazy."

But today the snow is gone and so is the wind, and now, with "a dab of breakfast" in her, Drue is feeling better. Maybe good enough to play solitaire like she usually does in the morning. Won't quit until she wins a hand, neither. Feisty, that's what she is when it's her against the cards. Stubborn, too. But a good kind of stubborn, the kind of stubborn you need to be when you're going nose-to-nose with nature and nature's trying to stack the deck against you. Yeah, stubborn is "The way a rancher should feel—to a point." The way Drue felt when the doctors wanted to amputate her whole foot, not just her toe, and she said, "Uh-uh. I don't want to have anybody have to take care of me. I didn't want to get along on crutches."

No, she wanted to be outside, working on the ranch that Karen calls "our big backyard." Having the satisfaction, like Ron says, "of working for yourself. Seeing the product of your labor" as you try to "come out on the positive side of the dollar."

Of course, "positive" is a relative term. Jiminy, just listen to Drue tell you about what happened "Oh, three-four years ago." The drought had hit pretty hard and the Hartongs had to lay out way more money than usual to feed the cattle, hauling hay across the state line, all the way from Farmington. At the end of the year, her accountant did all the arithmetic and, "He figured out I made a whole three dollars. Would you believe it?"

She laughs as she remembers the story—which just goes to show that her memory hasn't really gone bye-bye. Sure, parts of the past get a little blurry, maybe even misplaced for a awhile. But, doggone it, Drue'll find them sooner or later, haul them up to the present, dust them off, pass them around for others to share, and then take them back firmly into her grasp.

Some things, however, you never lose hold of. Like now, as Ron and Karen are getting ready to head up and tackle those five miles of busted fence. Drue walks with them as they leave the house, move through the yard, and prepare to get on their four-wheelers. In a voice that struggles to inject nonchalance into maternal concern, she says to her fifty-seven-year-old twins, "Remember—you call me when you're on the way back."

Then she turns and goes into her house. Maybe to feed those cats. Maybe to do some reading—provided the book isn't filled with sex, sex, sex. Maybe to think about getting up with Karen tomorrow to spray some of that nasty Canada thistle. Or maybe just to think of all the things that remind her that she's done everything she ever really wanted to do. Like practically living in a saddle. Like finding a good man, staying married to him for more than a half-century, and raising a fine bunch of kids. Like watching baby calves wobble onto their legs for the first time. Like going four-wheeling in some of the prettiest country nature has come up with, where the Little Navajo dances along and you don't have to look hard to find pieces of rainbow.

[Drue Hartong died Feb. 3, 2009.]

The old McBroom place

Chalk Mountains

VERMILLION RANCH

- *Moffat County*
- *Owners: Dickinson family*
- *First ranched by the Sparks family in 1885*

Vermillion Ranch is one of the most remote ranches in Colorado. It lies in the extreme northwestern corner of Colorado, and its public-land grazing leases extend into Wyoming and Utah. Its owned and leased Colorado lands span the distance between the Green River to the west and Vermillion basin to the east. Bottom lands along Talamantes Creek at ranch headquarters provide some hay for winter, but it's the vast Bureau of Land Management grazing leases that feed the cattle most of the year. This is a landscape that can be stark and lifeless or colorful and fecund depending upon the time of year. Land along the cliffs of the Green River in Browns Park boasts overstated scenery, while Vermillion Basin is more aloof. In between lies Cold Springs Mountain with its endless views and remarkable wildlife: elk, deer, sage grouse, and practically every other Colorado creature that prefers life away from humans. North Carolina immigrant Charlie Sparks got to this place in 1885. His marriage produced a daughter, Polly, who wedded one A. Wright "Dick" Dickinson. Dick and his children run the ranch today, and three grandchildren, Kate, Ira, and Ian, will most likely take over one day.

Pronghorn antelope

Bringing the cows down from Wyoming

Sage grouse

Elk herd on Cold Springs Mountain

Kate Dickinson, age eleven

Ira Dickinson, age eight

At the end of the road to the Middle of Nowhere lives a man who was born old. His name is T. Wright Dickinson, but don't bother asking what the T stands for because he won't tell you. He *will* tell you a lot of other things—a few of them personal, many of them polemical—once you get to his home. But before you can reach that home, you must drive long miles through a vast land of ruthless beauty, a land of biblical sunsets of purple and gold and primordial canyons that can swallow you up then spit you out and leave you feeling lost and found at the same time. A land that is hard and merciless, but so dramatically seductive that, even if you only linger, it just might invade your blood and your soul and never let go.

T. Wright's mother warned him about that. She warned him not to fall in love with the bleached mesas and the vermillion cliffs and the dirt that somehow manages to grow grass even when it's hard as rock and half-dying of thirst in the parched summers. She warned him not to fall for the land's tricks, for the whispered promise of endless space, for the smell of fresh-mown hay, for that first snap of cold that sneaks up on you in the fall and tastes like sweet electricity. She told him the ranching life was crazy, how it seemed everybody was against them—a government that was willing to renege on promises and breach contracts and laws, always wanting to tell them how to go about their business; self-righteous environmentalists who appointed themselves saviors but only knew how to put the land up on a pedestal. Even the very people they were trying to feed, fellow Americans, seemed to be against them, disapproving, unwilling to understand. Everything they had or ever would have came from "dirt, cattle and sweat" she told him; don't expect any favors, don't expect anyone or anything to cut you slack. Not out here.

But it didn't matter what his mother told him. T. Wright fell in love with the land and the life just the same. Just as his parents had. Just as his grandparents had. Just as his great-grandparents had. And so the man who was born old, who at the age of five took to responsibility the way a magnet takes in steel filings, grew up fast, leading the way for two sisters and a brother. And so they all grew up together, lived together, worked together. Hung tough together. Closed ranks together. Because that's what you do when your last name is Dickinson and you live on the unforgiving land at the end of the road to the Middle of Nowhere.

There he is, standing at the doorway to the house he grew up in, offering you his hand and a wary smile. He's long and rangy, looks taller than six-one, and has a face suspiciously boyish for a forty-six-year-old man. Opie of Mayberry turned wiser and harder. All that's missing is a cowlick, and even that will appear later in the afternoon and look absolutely in place. Despite his length, his stride is short and quick, almost like he's in a perpetual hurry to get somewhere and doesn't have time to merely amble.

Right now, those legs are carrying him to his truck. He's on his way to fetch the dump rake so all the extra scraps of hay can be collected. Hay can't be wasted, not out here where it seems the ground only grudgingly allows things to grow. Maybe that's why, as one friend says, "T. Wright fights for every blade of grass." Sometimes he's fighting nature; sometimes he's fighting the government. But whomever he's fighting, most people will tell you it's pretty hard to make T. Wright Dicksinson blink first.

It's also pretty hard to orchestrate an interview with him. Ask him a question about, say, what qualities a rancher needs to be successful, and you just might hear, "Now, forgive me, but I'm gonna shift the focus of our conversation rather substantially." And then, over the incessant ringing of the seat-belt alarm bell (he refuses to wear one), what you will hear is a lengthy explanation about why he thinks conservation easements are fine, but overrated as far as preserving the land from development. And during that explanation, you will hear him say that easements are "but a postage stamp of what the agricultural industry continues to conserve every day"—at least compared to what happens when the "decision makers of this landscape get up every morning and say, 'No, I'm not gonna sell that sumbuck, I'm gonna keep ranching.' *Those* are the people who conserve and protect this landscape." In other words, you don't have to sign a piece of paper to keep your land productive and agricultural, you just have to keep working it with as much elbow grease and gumption as you can.

With T. Wright, there's no getting away from this hitch-up-your-pants-and-just-get-on-with-the-job philosophy, and it's easy to see why. It was flat-out instilled in him. Talk to his father, seventy-seven-year-old A. Wright Dickinson—who pretty much everybody calls Dick—and you'll understand where the son learned the self-reliance mantra.

"Never had much use for psychologists," says Dick, a former president of the Colorado Cattlemen's Association and as plain-spoken a man as you'll find. "Whatever comes down the road in life, I've always said just look in the mirror, look at the problem and decide if you're gonna solve it. Ask yourself, 'Am I gonna handle it, or is it gonna handle me?' Then, once you make a decision, do it, and never look back."

It's just like being a rancher. Dick knows that "Some like us, some are agin' us, but we just do it. Wouldn't want to find out what it was like not to be ranching."

Dick'll look a man straight in the eye when he talks to him because, well, subtlety isn't his style. Out in the Middle of Nowhere things are often more black and white and you handle them the way you have to—directly. The land belonging to the Dickinson's Vermillion Ranch, and the sprawling miles around it, was once outlaw country—the infamous Tom Horn and the even more infamous Butch Cassidy were no strangers—and different generations of ranchers learned to be alert and ready to take care of their own. Livestock rustling sure wasn't unheard of back then, and, fact is, it still isn't today. Maybe that's why a sign in the Dickinson kitchen reads, "Trust Everyone But Brand Your Cattle!"

But T. Wright *doesn't* trust everyone. How could he? Not when there's people out there trying to cut him and his family off at the knees for reasons that are just plain wrong. Not when those people are busy fashioning nooses of red tape and "We've been subject to every environmental law and whim there is."

The truck has become a soapbox on wheels as it rolls past the strata of rock formations and scalloped green and brown bowls, ducking in and out of immense cloud shadows. It crosses into Wyoming, which is only six miles from his house, and is one of the three states—Colorado and Utah are the others—where the Dickinsons maintain ranching operations. And, by the way, exactly how much land and how much cattle they have is "none of your business."

The reply is delivered gently, with a smile, but firmly. Sure, it is considered bad manners to ask a rancher those questions; it's like asking someone how much money he's got in the bank. But many will overlook the rudeness of an outsider, smiling tolerantly at the breech of manners, and provide an answer anyway. Not T. Wright. "That's an impolite question and, no, I won't answer it," he says. Besides, there are more important issues to talk about.

Over the thrum of the truck's engine, T. Wright continues to address them. The topics may be old hat, but his passion seems fresh. Of course, it had better be because the threats he sees are always out there lurking. Take his ongoing battle with the government over who knows best how to manage the land, who knows best how to preserve its integrity and usefulness. The Dickinsons have been working the land since 1885 and have built enough ponds for livestock that "we're half beaver." Some of those ponds and dams are on public property because the Dickinsons hold grazing rights on thousands of acres of federal land, and there have been to-dos over whether those dams belong there. There have been vocal and jurisdictional skirmishes with the government about whether the family's cattle have violated and compromised wildlife sanctuaries. There have been arguments with environmental groups that don't think ranching—or any kind of development—is good for the land; that think it should be kept pristine and wild.

T. Wright has called some of these outside groups "radical revisionists," and resents the fact that they think they can come in here and know what's best for the land because "We were worried about these lands before those folks thought it was cool." His tone is a cross between a snarl and a left hook; just like his father, he's got no time for subtlety.

He pauses, but not for long. And when he speaks, his words and his sentences are loosely stitched together; sometimes they stumble into each other, but it doesn't matter because he's not about to come up for air until he's gotten to the guts of what's eating at him. Damn right he's convinced the Dickinsons are one of the big reasons the land is as healthy and viable as it is. Damn right "any sonofabitch that wants to come and take a pot-shot at me better understand that." They also better understand that "Right, wrong, or indifferent, for God and everybody to see, is there's 123 years of this family's stewardship and management of this area. And I think we can reasonably put our record against anybody else's, any place else, and say, 'I think we've done all right.' "

There's a good chance a lot of folks in Moffat County might say pretty much the same thing about themselves. Ag people out here are forged on a different anvil, one that maybe makes them a little harder, a little less likely to bend. Moffat County likes to call itself "The Real West" and whether you believe that or not, it certainly seems to be the heart of the Middle of Nowhere. It's about as northwest as you can go and still be in Colorado—but not by much, since the Utah and Wyoming borders effortlessly cast shadows here. The county has 4,756 square miles of ground, second most in the state, but only 15,000 people. Do the math and that works out to about 3 people per square mile, which is an awful lot of elbowroom. Another thing about Moffat County—and this rankles a lot of its citizens—more than half of it is comprised of public lands, land where the federal and state governments get to call most of the shots.

Big Government—and that means Washington, D.C. and Denver—has been a burr under the saddles of a lot of folks. But none of them feel the prickles more than ranchers. For ag people, government authority has meant navigating through the thick language of grazing permits, worrying about fees being raised or grazing lands being shrunk. It's meant that living in the Real West can be a real pain in the ass. And, if you ask T. Wright, it's been that way for a while, maybe since not too long after Charlie Sparks—his great-grandfather—arrived from back east and started raising sheep.

Other people came too, lured by the vastness, seduced by that ruthless beauty and the possibilities within all that space. They were homesteaders ready to sink roots into the hard ground, to pledge their sweat and good name in a partnership with the land. People with high hopes and uncommon persistence, men and women not given to giving in, not willing to waste time on despair or negativity. You stuck around long enough, you *had* to become an optimist; eventually, the land and the life would weed out all the nonbelievers.

But struggling with the land was one thing. Struggling against shifting philosophies was another.

"Families came out here into the West and invested lives and fortunes," says T. Wright, his tone getting flinty. "Then, subsequently, over time, the federal government and society has said, 'Oh, we want to re-think that contract, we want to do that differently,' not remembering the agreement that folks entered into when settlement was first encouraged."

Time went on and it seemed like the wide-open spaces got filled in quite a bit by laws, restrictions, and regulations. By the time the 20th century was waning, as T. Wright recalls, "People in this community were under assault from a whole host of environmental and governmental and other folks trying to tell us how we ought to live and work and what we ought to be able to do in our landscape. And that doesn't sit very well out here. There was a 'county movement' starting at the time. The feeling was county government should have a seat at the table, too. We should have a voice."

In 1994, thirty-two-year-old T. Wright Dickinson got a chance to act as that voice when he was elected county commissioner. Maybe thirty-two sounds a little young to some folks, but "you have to understand, all the issues we were talking about, I had been dealing with since my teens and twenties, when I come out of high school and college. Public lands, and dealing with the [federal] agencies and dealing with our trade organizations, hell, I was well-versed in all that."

Yep, he was ready all right. Hell, he'd probably been born ready. Just the way he'd been born old. Just the way he'd been born to fall in love with the land, the first member of the fourth generation of Dickinsons to be smitten.

Charlie Sparks came out of Traphill, N.C., pointed his nose toward the sunset and didn't stop moving until he arrived in the Real West in 1885. It may have been a no-man's land, with the brutal winters and stingy rainfall, but Charlie liked it well enough to stay and make his own pact with the country. To his way of thinking, the rugged land was best suited to raising sheep, so that's what he did. Got along all right with the cattlemen in the area, too; at least well enough that nobody tried to run him out.

Of course, if they had tried, they might have found themselves in trouble. Charlie was one of the best rifle shots around, good enough to match Tom Horn at the turkey shoots that were held down in Brown's Park. But his aim wasn't limited to bull's-eyes. He was an unsentimental man, had no problem getting rid of the weak. A horse got old or sick, couldn't make it through the winter on its own, Charlie would put it down with a bullet. That's the way you had to be out in the Middle of Nowhere.

If Charlie was tough and pragmatic he was also shrewd; shrewd enough to acquire more land every chance he got, shrewd enough to pass that tactic onto succeeding generations. Except for during the Depression, when no one had any money anyway, his people were always looking to add on to Vermillion Ranch. And, speaking of the Depression, that marked the end of another of Charlie's ventures. When the government had decided it wanted to reintroduce elk into the West, Charlie got himself a bull and two cows from up in Yellowstone Park. The day the animals arrived in Rock Springs it was a pretty big deal; even made the local papers. Then Charlie set about to growing a herd of elk. But when the damn Depression came, well, he was hurting just like everybody else. So when the county decided it was going to charge him personal property tax on that herd, Charlie said, "What elk?" opened the gate and let 'em go. Which is one reason why, eighty or so years later, an estimated 70,000 elk were roaming the areas of the White River and Black Mountain, probably the largest migrating elk herd in the world.

One of Charlie's daughters married the son of a mine superintendent. That union produced A. Wright Dickinson, who grew up on Vermillion Ranch. Eight years after Dick was born, Rock Springs, Wyoming, welcomed the birth of Pauline Radosevich, a descendant of Croatian immigrants who had come over to work the mines there. The Dickinson and Radosevich families were well acquainted: Pauline's grandfather had bought property from Dick's grandfather; Dick was the best man at the wedding of Pauline's cousin. Eventually, those eight years between them didn't matter much. They met socially at one of those dances at Lodore Hall down in Brown's Park and sparks started flying immediately. It was just "one of those things," is how Pauline would describe the event years later. It wasn't a long courtship — maybe nine months — "a few dances, a few rodeos and, y'know." Rock Springs and Vermillion Ranch were sixty-two miles apart, so it wasn't like the couple could just get together whenever they had a notion to. And "you can't waste too much time, y'know?"

They were married July 2, 1960. About fourteen months later, T. Wright was born. Barely a year later, Jeanne came long, followed not two years later by DeeDee. Dick and Pauline—everybody called her Polly—took a breather then and waited five years to have Marc. Although T. Wright was the one his mother liked to say was "born an old man. Even when he was little, he was responsible," pretty much all the kids were taught "the responsibility of ownership" early on. When Marc was five, for instance, and had to borrow money to buy some cows that he was going to raise on his own, he could only put an X on the documents.

If the Dickinson children were raised to be independent, there was a reason. You needed to be self-reliant in the Middle of Nowhere, where neighbors were scarce and the separation between people and services was vast. Where a round trip to the nearest supermarket meant at least 120 miles across that hard and beautiful land; across the chalky-white hills and the prairie pallet of pink and red and gold. Through the foreboding but magnificent Irish Canyon, whose sheer cliff walls could put the fear of God into you. Past the sagebrush and the Indian rice grass, and the stray herds of galloping antelope that could cover broad swaths of territory without ever passing a fence or a sign or anything else to suggest the presence of man.

Distance was always a given, although it was also relative. All the Dickinson kids went to the one-room Brown's Park School, which was only twenty miles from the ranch — or forty miles for the round trip. Or eighty miles if you figured in the extra back-and-forth to home and then school to pick up those kids. Yeah, you sure had to cover a fair piece of ground to reach a tiny place. Squatting out in the middle of all the land, the school looked even smaller than it was. But even close up it didn't exactly take your breath away. It could only accommodate fifteen students, although when T. Wright attended, there were never more than eight. The year he started school, there was just him and two girls.

Growing up for T. Wright was mostly a ranching saga. He didn't play sports or compete in rodeos like his other siblings. No, T. Wright loved working with the trucks and the tractors and the bulldozers, driving them, hands on the controls or inside the engines, probing and fixing. But he was more than just a whiz with machines. He was an FFA'er, a Future Farmer of America — that was his game, that's what he wanted to do. That's what all the Dickinson kids wanted to do. Out there on Vermillion Ranch, at the end of the road to the Middle of Nowhere, they were a tight band of agricultural conspirators, reinforced by their family's ranching tradition and sequestered from distractions. But the distances that kept the family's vision unified and made them close and loving also touched something deep inside T. Wright, murmuring a song that lured him to places where a man could be quiet and solitary and learn to prefer things that way.

Funny — well, maybe not — but when you first hear him say it, you don't know whether to believe it. A lot of people probably wouldn't, people who like him and people who don't. People who have experienced his outspoken performances as a Moffat County commissioner, as chairman of Great Outdoors Colorado, as a member of the boards of the Colorado Cattlemen's Association, or Club 20, the Western Slope lobbying group.

But there, in the dining room of his home, sitting at the table, waiting for his mother to create some of her typical culinary magic, T. Wright Dickinson is insisting, "I'm extremely introverted by nature; hardest thing I have in the world to do is stick my hand out and meet somebody new." You see, "We were raised in an adult environment without a whole lot of interaction with other kids. Yeah, we had cousins up the creek, but, well, meeting people is not an easy thing for me, it's a learned behavior on my part."

When you grow up the way he did and you're "wound tighter than a two-dollar watch" to boot, time alone is time treasured. Once, when T. Wright was in an ag leadership program, "they did one of those Meyers-Briggs temperament indicators and they found out I should be something like a heavy equipment operator. Those occupations that are solitary, that's where I'm comfortable at." That wasn't exactly a revelation for him; at one point in his life he'd thought about the solo life of a truck driver, something where he wouldn't have to "backslap or glad hand." Something for a man whose idea of recreation is to "go drive a bulldozer."

He looks up from his hands. "Now this one over here," he says, tilting his head toward Polly, "She doesn't know a stranger."

His mother smiles back, not pausing from her labors with the cutlets she's preparing now that the homemade tapioca pudding is cooling. She's proud of all her children, but, well, maybe there's something a little bit different about how she feels towards her oldest. After all, it was T. Wright who, after he'd been the family's go-to guy on the mechanical and building end of things — its "grease monkey" — rose to the occasion and became the family's "politician" as well. It was T. Wright who took the fight where it needed to be taken.

Just when the bad blood with the government started was hard to say, but things sure got to boiling furiously in the late 1990s, when the administration of President Bill Clinton declared the Vermillion Basin off limits to natural gas drilling. That really ticked off locals — including the Dickinsons — who had long been chafing at what they felt was government intrusion, and were eyeing the revenues that development could pour into local businesses and county coffers. On top of that, they scoffed at the notion that the public lands needed to be kept pristine and untrammeled. Hell, they'd been used for more than a century by generations of ranchers like the Dickinsons, those half beavers, who had been working the land, building livestock reservoirs — sometimes with teams of horses and slips — and grazing their cattle there. And those cattle grazed on a lot of land, too. One newspaper account had it that Vermillion Ranch leased over 160,000 acres from the U.S. Bureau of Land Management in Colorado and Utah alone.

In 2001, some of that territory was threatened when the government tried to turn 6,000 acres of grazing land leased by the Dickinsons over to the Brown's Park National Wildlife Refuge. The refuge was run by the U.S. Fish and Wildlife Service, whose tighter usage restrictions didn't figure to be very welcoming to grazing cattle. "A land grab of the highest order," is what T. Wright called it. "Citizens all over our county are going to be outraged by this."

At least the Dickinsons sure were. Soon, they were embroiled in legal skirmishes over the fact that the government wanted them to dismantle two reservoirs it claimed had been put up illegally. For their part, the Dickinsons thought the government was reneging on a longstanding working agreement. Or at least acting in bad faith with a family that had been hardworking stewards of the land. Besides, they maintained, rather than being a threat to endangered species of wildlife, grazing cattle helped control noxious weeds and actually improved wildlife habitats.

By then, T. Wright was into his second term as Moffat County Commissioner and, depending on your point of view, his take-no-prisoners style was admirably tough or just-plain intransigent when it came to the issue of public land usage. To a lot of locals, he was practically a knight-errant, someone who shared Moffat County resentment over what they saw as increased federal efforts to further limit local access to lands in the county and was willing to go fight any bureaucratic dragons on behalf of his beliefs. And it wasn't just the Brown's Park National Wildlife Refuge that was a sore spot. Why were there all those suffocating restrictions on landing hang gliders in Dinosaur National Monument? Why was motorized travel in the Moose Mountain area banned by the feds? T. Wright became the point man for a lot of people who, like him, were sick of the government and those "radical revisionists" dictating everything that happened on Moffat County land. After all, wasn't it T. Wright and people like him who had worried about the land and cared for it before all these outsiders suddenly found it so trendy to become its so-called protectors? Damn right it was. Damn right T. Wright knew what he was talking about when he said, "Selfishly or populistically on my part, I believe the folks that manage the land ought to be the same folks who work the land."

The thing about T. Wright was, he spoke the language of the county. Sometimes he spoke it eloquently, sometimes he spoke it angrily. But he always spoke it with the kind of sagebrush sensibility and accent that a lot of people from the Real West recognized and felt comfortable with. As one rancher put it, "T. Wright? He's the real deal. He's as real as it gets."

In 2003, Moffat County struck back against the government and radical revisionists — the forces who, everywhere you looked, seemed to be trying to siphon off control of the land, taking it away from the people who knew best. What the county did was invoke an obscure 1866 U.S. mining law that gave it the right to manage some 1,000 acres of roads that threaded through public lands, many of which were protected by a federal wilderness designation. Although the law had been repealed in 1976, there was a loophole: states and counties were still entitled to claim right-of-way control if they could prove that the route had been constructed before the land had been reserved for wilderness. T. Wright's second term as commissioner had ended several days before Moffat County filed its claim, but there was little doubt that he had played a significant role in the county's action. After all, the process to evaluate which roads to claim sway over had been a three-year enterprise.

"We have done nothing more than assure the county citizens and the American public in general their right to continue to travel on our public lands," said now-private citizen T. Wright Dickinson, once again willing to step out from behind his shyness and speak up. As you might expect, his outspokenness made him a familiar name in newspaper stories, which referred to him as a "politically connected cowboy" and a member of "one of western Colorado's most powerful clans."

Of course, his profile hadn't exactly been diminished by his appointment in 1999 to the board of Great Outdoors Colorado by then-Governor Bill Owens. GOCO—as it's called—was created by a statewide referendum to serve essentially as a trust fund that would invest millions of dollars in Colorado Lottery proceeds to protect the state's open space, parks and wildlife, principally by purchasing tracts of land and protecting them from development. In 2003, T. Wright was made chairman of the GOCO board, a move that didn't sit well with many environmentalists who already felt threatened by what they perceived as GOCO's shift away from purchasing land for preservation purposes and towards using its money for programs and purposes not oriented toward conservation.

Mindful of T. Wright's role in Moffat County's attempt to gain control of roads in public lands, one former GOCO board member publicly referred to the chairman as "the primary progenitor of the controversial idea to allow new roads to be built throughout Colorado's national parks, monuments, wilderness areas and other natural lands." The board member also claimed that Owens was "thoughtless in choosing who is in charge of protecting our remaining wild lands and open space." The fact that, at the time of his appointment, T. Wright was serving as chairman of the board of Club 20 didn't exactly endear him to environmentalists, many of whom saw him as some kind of prairie pirate intent on plundering the land. After all, Club 20, a Western Slope booster and lobbying group made up largely of developers, ranchers, farmers, miners and energy executives, was not known as a champion for the cause of public land ownership.

To T. Wright, the criticism made sense—considering it was coming from a lot of those radical revisionists. To his way of thinking, GOCO wasn't solely intended to be a land acquisition body; preservation of state parks and species protections, for instance, were part of its mandate, too. So he wasn't surprised by the hard feelings; he knew about the chasm that separated the different sides. As he would later say, "It is tremendously disappointing and frustrating to me that the very folks within our society that have a tremendous and deep-seated value of these landscapes continue to be so distrustful and setting up in opposition to each other when we have so much in common. By that I mean the environmental folks and the folks that want the land to have meaning for us.

"And, see, that's the passion for T. Wright, that's what gets me out of bed every morning—articulating and recognizing that I have a responsibility to help my folks get their point across, and also at the same time, reaching out to those other folks that are willing to meet us halfway."

He didn't know how many people were willing to do that. What he did know was that "There are three kinds of people in life—those that build bridges, those that burn bridges, and those that stand in the middle of 'em." Depending on your point of view, he might have a hammer in his hand, or a lit match, but he sure wasn't going to be twiddling his thumbs all wishy-washy like. You see, "I got politically active because my folks, my community, my industry were getting run over and not getting a fair shake to my way of thinking. People in this community, ranchers, were under assault from a whole host of environmental and other folks trying to tell us how we ought to live and work and what we ought to be able to do in our landscape, and that doesn't sit very well. So I stepped up and said, 'I don't think this is right.' "

When it came time for him to exit public office, T. Wright was more than okay with that. Or, as he put it, "Thank god for term limits." He was done with politics for good; he'd had enough of being an elected official. After all, getting into it, he "clearly didn't understand what it meant to be an elected official. But I guess if anybody ever told you about what the responsibilities for the job are and what it entails, then no one would ever run for it."

But if he was happy to go back to being more of a grease monkey than a politician, he had no intention of becoming one of those people content to stand in the middle of the bridge.

T. Wright Dickinson has been asked a question about what he likes least about his profession. He pauses for a long time. Then, not surprisingly, he says, "I won't tell you what I like least about it, but I'll tell you about what I find its greatest irony."

His tone slides toward brittle, and he starts to speak faster, words tumbling out of his mouth, some of them trampled by the ones coming up behind them, the way they do when he's about to mount his soap box and have his say on something he feels strongly about.

"The public, continually, when surveyed has always rated ranchers and farmers, held them in a very high esteem, and yet every paper article, stuff that comes out, it's 'big ranchers this,' or 'somebody's doing that to destroy the environment.'" A quick breath and, "There's this disconnect between, well, y'see the public perceives ranchers as good, hard-working folks and yet vilifies us, or certain elements continue to vilify us, and yet we're the very folks who have continued to steward and sustain this land. I mean we cared about this land before anyone else did because it made our livelihood, and then for someone to come along and say, 'You ain't doing it right,' without even setting down and having a conversation about, 'Hey, we want to understand how you see this before we say what we think,' well, what's the point of that?"

A lot of them radicals might not know it but whether you have "one acre or 10,000 acres," what drives anybody who works the land out here is "stewardship and betterment of the land. That's why we go to bed at night, that's what gets us up every morning." Out here, "it's all about the connectivity of place—that's what ranchers are about. This is our place. It's where we have chosen to make our stand. We may bend, but we won't break."

And after years—no, decades—of working the land and becoming intimate with it, he's flat-out pissed off that his family *still* has to battle with government and outsiders. At least that's what it sounds like when he says, "It's taken all of our family's persistence, stubbornness, patience, political clout, whatever, to prove that we were right and that we actually knew what the hell we were talking about and that we weren't out there to rape, rob and pillage."

And the damndest thing is that there are still people out there who doubt the Dickinsons—at least if you ask the Dickinsons. As Polly says, "We wonder why we're disliked by some of the public. All we're trying to do is raise food—where did that get lost in the public's mentality? We know we're at the mercy of nature, we know we have to fight the elements, but why do we have to fight fellow Americans?"

Polly resents those newspaper articles that refer to the Dickinsons as "one of western Colorado's most powerful clans," cracking, "I wish someone had told me we're so powerful." Similarly, T. Wright remembers one editorial that "made me sound unwashed and unworthy." But does he think he's controversial? "No, not at all." You wait for him to smile when he says that, but he doesn't. Instead, he's busy insisting that, as far as being called the "primary progenitor" of plans to invade wilderness areas with roads and vehicles, well, "I'm not even sure what that means."

Of course, don't believe him for a minute. Just forget the grease monkey-simple country-boy mantra, the simplistic self-description that goes "I'm somewhat reasonable, I'm somewhat modest and I can actually maybe think a little bit." And don't pay all that much attention to the fact that he didn't get around to graduating from Colorado State University ("I took the two-year plan"), either. See, when you get right down to it, T. Wright is not only "real," he's real intelligent. Spend just part of a day with him, and you might hear him talking about the book *The Whale and the Supercomputer*, expounding on the subject of "superior barbarism" and using the word "heretofore" more than once. So, you bet he knows what any unflattering descriptions means—be it "bully" or "primary progenitor." What's more, he's not immune to name-calling because "Of course, anybody would care. We're all humans. On one level you have to shake that off, but no one likes being maligned."

Maybe not, but when it happens you just have to cowboy up. Let the criticism roll off your back, but expect more of it and be ready. Keep smiling, but maybe have one fist half-clenched behind your back. Y'know, just in case. Maybe keep in mind what's written on the small plaque that's back home, hanging on a wall near the dining table, not far from photos of Pope John Paul II and George and Laura Bush and the piles of paper and envelopes that are scattered everywhere in a house that, like its inhabitants, doesn't put on any airs. The plaque bears an "old Celtic blessing" that goes:

May those who
Love us, love us
And those that
Don't love us
May God turn
Their hearts
And if He doesn't
Turn their hearts
May He turn
Their ankles
So we'll know them
By their limping

It's defiant and amusing at the same time, which is half of what Polly sounds like when she says, "It gets very discouraging, sometimes. But I guess it's the pioneer spirit in us that keeps us fighting."

Actually, it's the pioneer spirit *and* a tenacious conviction about the rightness and maybe even the righteousness of their calling. You see, agriculture is a noble profession; it's the damn backbone of this country. And if you don't believe T. Wright Dickinson, well, maybe you'll believe Thomas Jefferson.

"Jefferson was so right," says T. Wright, voice almost trembling now as he paraphrases one of America's Founding Fathers. "You let your cities fall and the farms will spring up again. You let your farms fall and there's never been one civilization that has ever stood when it couldn't feed itself. Not one!"

Not that helping to feed his country is the only reason T. Wright gets up every morning and decides not to sell that sumbuck. Hell, ranching isn't only about altruism and a sense of duty. Or, as he puts it, waving his hand at the landscape that's unfurling past the truck, "Some folks come along and say, 'Gee, isn't this pretty?' Well, we admire the beauty and all that kind of stuff just like everybody else, but that's also my producing landscape. That's the way I make my livelihood. And if somebody says, 'Well, that's not important to us,' well, I got a mortgage that says that's important to *me*. That's my 401(k), right there."

But it's more than that because "Some folks look at this, and to them, this is wilderness. But to me this is home." And because it's home, he's utterly comfortable in it; a man plopping down into an old easy chair in his living room, dropping his guard, knowing he's safe. You can hear a calm relief in his voice when he stops the truck, jumps out and takes a few of those jackrabbit-quick strides over some spongy ground and launches into a short, concise lesson on this "mixin' bowl of geology and ecology." You can hear it when he's talking effortlessly about prairie grasses, like sagebrush ("volatile oils that inhibit digestion"), rabbit brush ("not too palatable"), white sage ("we also call it winter fat; it's highly palatable for cows and sheep"), and hop sage ("sheep like it"). He may not be "some great plant taxonomist," but he knows his stuff, he knows that "so much of this range thing is as much an art as a science." And if you comment on how much he knows about the flora, he sidesteps the compliment with "Every profession has its little nuances, stuff we can talk about so we can act like big shots."

When he utters those last few sentences, you can almost see the fists starting to clench a little bit, feel the business-like T. Wright reappearing. Like he can't let down his guard for too long and he needs to go back to being blunt. But even he isn't immune to certain moments, moments when his voice floats beyond the realm of respect for, or comfort with, the land and reaches a place where all the sharp edges melt way; the kind of place a solitary man probably doesn't share with too many folks. A place like where he is now — standing on Pine Mountain and telling you, "That's the Uintah Mountains. On a clear day, you can see from here, oh, about back to Evanston, Wyoming. And behind us, well, it's too hazy today, but you can see the Wind Rivers in Wyoming. That'd be about 100 miles."

The wind is gusting, blowing clean and loud, and over its noise, T. Wright is looking down from 9,200 feet and talking about hearing how "Some old guys in the 1920s and '30s told my dad about going on top of this mountain and gathering wild horses down to a 'cral. They used to run them horses off here full-tilt, never let up." His grin is starting to stretch his face as he says with wonder and admiration, "Can you imagine going off of this goldang rock pile? It'd be suicide!"

Suddenly, he's not a politically connected cowboy drawing a line in the sand against invasive government policy and radical environmentalists. He's not a sober businessman talking about his investment, or a veteran rancher expounding on the nuances of grass. Suddenly, his tone is more gee-whiz, his cowlick has appeared and his dark eyes soften as he takes in the land that splays out before him, green and undulating and going on forever.

The wind is still rolling across the land like surf on the ocean, carrying away the rest of his words — but not the emotions, not the *happiness*, behind them. They follow him back to the truck, lingering as he starts the engine and begins driving, not evanescing until he reaches the dump rake. He gets out, puts on his grease monkey hat, and commences to figuring out how to overcome two flat tires on the trailer. Soon enough, he's got it hitched to the back of the truck and he's hauling it back to his house. He's hauling it back because he can't waste a scrap of hay in a land that's not big on granting second chances. But he accepts that, just like he accepts so many other realities.

No, T. Wright Dickinson didn't listen to his mother when she told him not to fall in love with the land. But that doesn't mean his love is unconditional. You can't surrender all of yourself to this land, and you sure can't always turn the other cheek, either. Charlie Sparks knew that and so does his great grandson. Respect the land, feel that connectivity to it, but don't trust it. Because things are different in this place where the sunsets are biblical, the distances are immense and sometimes it seems the ground is only good for growing a chip on your shoulder. Yeah, you might rest easy at night after another long day of working this sumbuck, where everything you have comes from dirt, cattle and sweat. But you better not sleep *too* soundly; maybe you better even sleep with one eye open, staying alert, ready for anything nature or man can hit you with. But if you have to do that, then so be it. Nobody ever said this life was easy. Nobody ever said you got a free ride out here.

Out here, at the end of the road to the Middle of Nowhere.

Cold Springs Mountain

Green River sunset

Vermillion Basin

MANTLE RANCH

- *Moffat County*
- *Owner: Jim Carollo*
- *First ranched by the Mantle family in 1918*

The Mantle Ranch is just as remote as Vermillion Ranch. It is an inholding in northwest Colorado's Dinosaur National Monument. In addition to being remote, it is surrounded on all sides by the thousand-foot-high sandstone walls of the Yampa River Canyon. This serpentine canyon is a favorite of rafters and kayakers, who pass right by the ranch on their way through the monument, but probably have no idea about its remarkable history.

Charlie Mantle had just returned from World War I and was looking for a place as far from anywhere as possible when he found the Yampa River Canyon in 1918. He, his wife, and their five children lived as isolated an existence as they could. Trips to pick up mail in Maybell, forty hard miles to the east, or to get supplies in Vernal, Utah, an even harder and farther ride, were infrequent, so the family grew most of what they ate. The children were sent on the two-day horseback journey to get the mail when they were as young as eight. In fact, they grew up pretty much like American Indians, running around naked much of the time on the sandy shores of the Yampa and beyond! Nevertheless, each of them ended up with a college education. The Mantle family gave up their remarkable heritage when they sold the ranch to Jim Carollo in 2004.

Indian rock art

Indian rock art on the ranch road

Yampa River Canyon, ranch on right

WEST SLOPE

RUSSELL RANCH

- *Rio Blanco County*
- *Owners: Rogers family*
- *First ranched in the 1890s*

Russell Ranch lies along the banks of the White River, three miles east of the town of Meeker. The White River originates in the Flat Tops Wilderness and winds its way west until meeting the Green River in Utah. For this journey of more than one hundred miles, it is essentially natural, coursing pristine public lands of the White River National Forest as well as bucolic ranch meadows. It bisects the lower portion of Russell Ranch and its beautiful hay meadows. Another parcel of Russell Ranch lies farther west at a higher elevation in aspen, spruce, and fir forests above the hamlet of Buford. Russell Ranch got its start in 1907 when valley land was acquired by Robert and Nellie Russell. Son George Russell expanded the ranch. His daughter Marge married valley newcomer Allan Rogers, and when George died the Rogers' took over. Today, their son Ben Rogers operates the ranch.

White River

He's wearing a baseball cap and riding in a high-tech tractor, watching a computerized screen detail the precise diameter down to the inch of the hay bales it's forming and spitting out, but over the cool whoosh of the air conditioning his voice is full of a sweet reverence and respect for the time when building haystacks by hand under a molten sun was "an art form."

He's got a handshake that could fuse coal into diamonds, but a heart so soft he'll interrupt a chore to get out of his truck and cut a sprig of wild honeysuckle to bring back to his six-year-old son.

He's been practicing the delicate science of artificial insemination on his cattle for years and years, but he does all his inventory computations by hand in a notebook ("I'm in the stone age") and when it comes time to sell his calves he seals the deal on nothing but a handshake.

He loves biscuits and gravy at the downtown coffee shop, and cooks up a barbecue that could make a vegetarian convert, but he'll order the yellow-fin tuna steak (medium rare) at the Bistro on Park Avenue in downtown Meeker.

He dismisses the pop-culture notion of The Rancher ("some kind of nostalgic, *Bonanza* figure on TV") as something that's "not real and never was," and wears Nike sneakers instead of boots and "my buckeroo outfit" when he's working in the field, but still insists there are "certain rules, codes and etiquettes and things" that a "real western person knows," because "anybody who walks in the room with his hat on and leaves it on, well, I have to ask myself if he really is a cowboy."

He's been known to work through the soft summer nights baling hay and through the freezing winter nights tending to cattle, but there are plenty of times when, just on impulse, he'll stop whatever he's doing to "look at all this splendor in front of me." Times when he gets off his horse to check out "these flowers growing all around with these vibrant little purple cups on 'em," thinking, "Man, that's pretty."

He's got a birth certificate that says he's 42, looks ten years younger, and is married to a woman who is six years younger than *that*.

He's the son, grandson, and great-grandson of school teachers and will tell you more than a few times "I'm not the smartest guy in the world," but then talk about the time he held his own testifying before a U.S. Senate committee on the issue of how ranchers are an "endangered species."

Oh, yeah, he's a bit of an ambling contradiction, this fellow Ben Rogers is; someone who's "intense and focused" but "kinda easygoing." A guy who has plenty of self-confidence but isn't short on self-doubt, either. Someone who worries about a lot of things but is so comfortable in his own skin that sometimes it seems he could just go and take a nap there. A guy who's old school and new school; one foot in the past, one foot in the present, but both somehow twitching toward the future.

A jigsaw puzzle is what you might call him, except some of the pieces don't seem to fit together all that smoothly. At least not yet. Maybe they will someday. Maybe they won't. Hard to say for sure. But there is one thing you can pretty much say for certain about this fifth-generation son of the White River Valley who's part rancher, part philosopher, part jock, and part poet.

He's a one-hundred-percent interesting work in progress.

Some guys may like Massey Ferguson. Some may fancy New Holland. But Ben Rogers is a John Deere man all the way. Not sure why. Just is and always has been. He tells you this in his usual offhand-but-earnest way as he sits inside a John Deere tractor that is turning windrows of hay into golden, 1,100-pound bales, carrying him and a visitor along on a wave of noise and purpose.

Outside the cab's cool comfort, it's summertime sticky. The pewter sky is playing some kind of peek-a-boo game—will it rain or won't it?—and Rogers keeps a close watch. He wants to bale as much as he can this afternoon because if it starts to raining heavy he'll have to stop. You can't bale wet hay, he explains. A little bit of afternoon moisture, a little bit of morning dew? That's good, that makes the bale tighter, not so flaky. But an overload of moisture? Forget it. Then you gotta wait until it all dries out, and if it takes too long for that to happen, you probably lose some of the nutrients in the hay, and Ben Rogers wants to produce the best hay he can so he can produce the best cows he can. Because pride in his labor is "just something that's bred into me, I guess."

But if he's getting a little edgy looking at the sky, he's still acting pretty cool because, well, what choice does he have when it comes to the elements? "Hell, weather?" he says, rubbing his silver-flecked, dark-brown goatee that somehow makes him look more boyish than mature. "It's kinda one of those deals where you can't get pissed off at Mother Nature or what she's gonna do to you. You just deal with it, y'know?"

Chances are, if the weather holds, after supper he'll just bale tonight, alone up on the mesa, under cool moonlight, sitting in the cab and listening to his satellite radio play rock and roll oldies. Or maybe some talk show. Or maybe he'll just sit and think. Up in the tractor all by himself he can think of a "million things." He might think about ranching, about how he loves it. How "I'm spoiled, pretty much being my own boss." How, "I don't know if I could take orders from someone else."

He might think about how ranching is such damn hard work, how you're always worrying about this or that. About, "What's the corn market doing that affects the price of the grain you feed the cow? What's the green movement doing in the political makeup that's gonna drive those costs up that actually drive everybody's costs up? What's gonna happen in the fall when you go to your bank and say, 'Well, are we gonna get another operating loan for next year or are we done?' "

Maybe he'll think how sometimes it seems the government is "just losing sight of common sense." How it sometimes hires people that have "got an agenda and they like to twist and turn how they look at the data and the information so they can make it kinda tough on you." How they make you "fill out about 15 million forms for each little thing and it's just ridiculous." And, sure, maybe ranchers don't document everything as well as they should but, *come on*, why does your own government have to act like you're the enemy?

And speaking of the enemy, maybe he'll think about some of those so-called environmentalists who think ranchers are just a group of "spoiled, rich, arrogant people who run the land, y'know, like we don't care about anybody else, or anything that's there, we just care about ourselves and what we're doing, and that is so far from the truth." There's a lot of those people, people who "think they know everything about everything that has to do with nature," people "who want to reintroduce all the wolves and put the forest back the way it was 100 years ago." That's crazy, stupid, because "the way I look at it, man is a piece of evolution, too, isn't he? The things he's brought to the table are part of the big picture."

Within the cocoon of tractor noise and radio babble he might just think about the land and how "I never take it for granted." Not its bounty of grass, nor the cows that live off that grass and have fed him and his family for five—going on six—generations in this place called the White River Valley. It is a place of simple beauty and grace. Y'see, unlike the overwhelming vistas found in other mountain valleys, the White River is less demonstrative, less showy and more modest, like the naturally pretty girl who doesn't have to wear a lot of makeup, but still melts your heart without even trying.

Then again, Ben Rogers might think about Cole, his six-year-old boy, or Charlie Ann, his baby daughter, and how he wants them to grow up on a ranch because that kind of upbringing makes you "tough and strong," gives you "good grounding" for anything in life. It gives you "integrity and values," and "maybe one of the problems with this country these days is we're getting away from that."

Cole and Charlie Ann's father might also wonder if they'll grow up wanting to stay on the land the way he wanted to, or whether they'll go off somewhere else and he'll have to decide what to do with the ranch. Maybe take out easements to protect it for agriculture. Maybe sell some of it off, which would break his mother's heart—and his, too—because the last thing he wants to do is see more of those developments springing up, the ones with million-and-a-half dollar home sites where "you can sit around and enjoy those manmade ponds and manmade landscaping and manmade structures and I guess say you're living your piece of the West."

God, he'd hate for that to happen. Hate it for a ton of reasons. Hate it because of the debt he feels to the people who came before him and the responsibility he feels to the ones who will be here after him. But thinking about things he'd hate to happen is something he'd rather just push right out of his mind. Which is easy enough to do. Especially at night when the air is cool and just moist enough to make perfect bales. When the light is silvery and he can look out at the tractor's moon shadow sliding across the hayfield, like a finger moving across the pages of a book, a book he's both reading and writing. A book someone else started and he doesn't want to end.

The ranch that would grow to 1,280 acres in some of the prettiest country in Colorado got started in 1907 by Robert and Nellie Russell, which was five years after they first laid eyes on the White River Valley and began putting away money to buy their own piece of ground. Robert and Nellie had a little boy, George, who grew up to be the visionary behind the Russell Ranch. It was George who, little by little, started buying tracts of land, cobbling together acres that were adjoined, so that instead of a piece of land here and a piece of land there interrupted by other properties, the Russell Ranch began to spread smoothly and contiguously. Then, slowly, he purchased land higher in the mountains, a 160-acre plot here, then another one next to it, then a third one next to that, building the Russell Cow Camp; buying land from homesteaders who were tired of the winters and the hardscrabble life and were ready to chuck the ranching along with the stills they had built to make a little moonshine. Pieces of those old stills would lie scattered on the ground for decades, like rusted bones off an abandoned skeleton.

Meanwhile, George Russell kept adding to his government allotment on the higher ground, obtaining permits for land that his cattle could graze on in the summer months, right under the gaze of Burro Mountain and Sleepy Cat Peak. By the time he was finished, he had permits for over 12,000 acres of government land; room for his cattle to meander and devour the Idaho fescue and countless varieties of other grass, growing healthy and meaty. The Seven Lakes Allotment and the land he bought later on were stitched together over long years, not carelessly but carefully. George Russell wanted to make sure he had a place for the cattle to move to easily after it was time to leave the summer ground. Then he wanted to be able to segue his animals back down to his home ranch for winter. See, some people just ranch. And some people ranch with a vision.

Of course, expansion didn't happen overnight. George Russell didn't much care for debt; wouldn't buy any new land until he'd paid off what he owed on what he had. Besides, it was hard enough making a go of it on what ranch there was. During the Depression, he and Mary, his wife, were squeezing their money so tight it practically bled. Some years, cows were going for no more than twelve bucks a head. Thank goodness for the fishing cabins they had built near the ranch, right off the White River. The dollar-a-night rental fee certainly helped. Then again, any extra income on the ranch helped. Always would.

George Russell labored on, expanding his holdings, working the land tirelessly. "There was no time to doodle-dawdle," his daughter Marge would recall, remembering the way her father would hardly take time to sip water from the canvas bag he carried with him, remembering how he could wield a pitchfork like he was a cross between Michelangelo and Frank Lloyd Wright, building haystacks that defied the wind, haystacks that rose twenty-five feet, so tall they "touched the sky."

Marge grew up on the ranch, loved its beautiful land and animals. She helped whenever she could, too. That is, when she wasn't going to school in Meeker, riding on her horse, or studying diligently. Just like her mother and grandmother, Marge was going to be a schoolteacher. Eventually, she was also going to be Allan Rogers' wife.

He had grown up in South Park, scion of a ranching family, a damn good rider and roper who competed in rodeo some. Then the U.S. Army made him one of those offers you can't refuse. He was heading towards Japan to be part of the invasion force when President Truman dropped the atomic bomb. Next thing Allan knew, he was part of the military escort bringing the Emperor Hirohito to Tokyo to sign the terms of surrender that ended World War II. The rest of Allan's hitch in the army was highlighted by boxing. He wasn't very big, but that 160-pound farm boy was strong and determined enough to finish undefeated as a G.I., good enough to win his division's middleweight championship before returning home.

Not long after he got back to South Park, he and his father decided to sell their ranch and look elsewhere. In his case, elsewhere turned out to be Meeker. Although they liked the land instantly and set about working their newly acquired K-Bar-T Ranch, Allan soon discovered there was a sort-of initiation in the valley. One night, at a barn dance, some of the local boys — maybe fortified by a little of that high-country moonshine — decided the newcomer wasn't all that tough. Decided to see what he was made of. Bad move. Especially for the one who wound up on the floor under the jukebox, out cold.

At some point, Allan met Marge. Wedding bells. She taught school and helped on his ranch. He worked the K-Bar-T and then sold it around 1970. Became a brand inspector for the state. Figured he'd do that some, maybe some rodeo, too. Then, two years later, Marge's father died and it was time for his son-in-law to take over the Russell Ranch. He did a good job, too. And he did it on his own terms.

See, Allan could be ornery, maybe even a little pugnacious at times. Definitely what you'd call straightforward. Thing was, there was no bullshit about him; he either liked you or he didn't like you and he formed an opinion pretty quick. And if he didn't like you, he wouldn't have anything to do with you. It was that simple. So was his work ethic. He was always on the go, couldn't stand still; a coiled spring of a man powered by nervous energy, always had to be doing something. Working outside was his drug, and so was relying on his own two hands and legs. He loved to walk — oh, he was a walking machine; bustling everywhere every chance he got. Sure, as ranches became more and more mechanized, he went along — to a point. One thing he never understood was the newfangled love affair for the four-wheeler. As a whole new generation of ranchers became addicted to the all-purpose vehicles, he stood back and shook his head. Never owned one. Hell, never even rode one. Give him a truck or an old tractor without a cab or, even better, a horse. He'd look at the new tractors, with all their fancy air conditioning and gizmos for baling hay, and say, "Aw, why don't you get out there with a team of horses and do it that way?"

Still, it wasn't like he was this Cro-Magnon rancher marooned in the past, treating change like it was some kind of toe fungus. He often told Ben, "Don't ever keep your head in the sand." *He* sure didn't. Take cattle. Allan was one of the first in the valley to have black Angus. Sure, he'd raised Herefords most of his life, just like everybody else. But he wanted to breed the best cattle he could, get the best price he could. That being the case, "I'm gonna raise what's selling the best and I don't care if they have polka dots on 'em!"

Ben listened real good because growing up, well, his dad was his hero. Those big shoulders, all that non-stop energy — *man, how does he do it?* Haying, fencing, working together, the boy watched the man and learned. And as the son got older, the father made sure to include him in decisions. Some ranchers, they couldn't let go, couldn't bring themselves to listen to their children. Allan was different. Oh, sure, he could be pigheaded; over the years he and Ben would clash more and more about the best ways to raise cattle, but maybe that was because the father taught the son early on that his opinion mattered. Like when they started going to bull sales together — Ben was probably in junior high — and the father would ask the son, "Well, what do you think?" Sometimes those sales were all the way in Calgary. Allan would take Ben out of school for a few days and they'd go up together. Make decisions together. Good memories for both, but especially for the son. Part of the glue that held him to his roots.

Roots that ran so deep you couldn't tear 'em apart with a backhoe.

If Meeker wasn't quite Mayberry, there were times when it may have seemed pretty close, right down to the fact that, as one native put it, "We were fifteen years behind the rest of the country in just about everything." But back in the 1960s and 70s, well, there just wasn't any better town to grow up in than Meeker. And there wasn't any place around Meeker better to grow up on than the Russell Ranch, in the heart of the White River Valley.

Ben had an older half-brother, Lance, but they were nearly two decades apart in age, so he grew up more like an only child. Maybe that's why from the time he came into this world — right there in Pioneer Hospital in Meeker on April 21, 1966 — he tended to be a little bit shy; a little bit cautious, someone who didn't mind keeping to himself. He wasn't the type to jump into things; first he'd sort of observe the situation, then decide. At least that's what his mother noticed about him. To his friends, well, Ben sometimes seemed wound pretty tight. Kinda like his old man.

Of course, all it took to loosen up either father or son was being outdoors with some time on their hands. Take learning to fly-fish — just him and his dad up there along a spot by Elk Crick, where there was nothing around but this old ranch house and 'crals. One day that spot would belong to folks who didn't know one end of a horse from the other. Then, instead of natural splendor, there'd be a trophy lodge and a trophy lake sitting in the middle of it and Ben would talk about how "I don't feel much welcomeness anymore; I'm not sure if that's even a word, but that's how I kinda feel." But that would come later. Back when he was a boy, the mountains belonged to him and his dad. And, of course, him and Patch.

Patch was Ben's pony. Learned to ride him when he was six. At first, he was mostly interested in chasing frogs aboard Patch, but then he got to be older and he'd ride after cattle with his father and grandfather through the whispering aspens and along Oak Ridge where the trees turned flaming red in the fall and there wasn't anything prettier in the whole world. One time, young Ben was riding with them up near the summer allotment, and there was this spring that came out at the upper end of the trail, below Burro Mountain, and, of course he couldn't wait to get there and get a drink. Every so often after that, when their rides would take them near the spring, there'd go Ben, making sure Patch got him to the water before anybody so he could take the first gulp of icy-delicious water. So, being that it didn't have a name to anyone's knowledge, his grandfather just started calling it "Ben's Spring." And then his dad started calling it "Ben's Spring." And pretty soon everybody in the family started calling it that. And even though you'd never find it marked on any map, inside the boy's head and heart, that was *his* spring and would stay *his* spring forever, even after he became a grown man. *Ben's Spring.* It made him feel tied even tighter to the land; like those homesteaders who had broken their backs to clear rocks and dig ditches and string wire across fence posts they had to cut with axes because chainsaws and gasoline-powered motors came along way too late to help them.

It didn't take a genius to see how Ben loved ranching, especially when it came to livestock. He loved the cattle; practically inhaled any information about them. Learned from his dad, and learned by being a diehard 4H member. Raised pure-bred Herefords and showed them at the Colorado State Fair and other shows, winning awards that his mother would keep documented in a scrapbook. One year, he even represented the whole state at the National Junior Hereford Association competition. When he wasn't showing cattle, he was learning to judge them. Livestock judging was serious business; you competed on teams. And like any team, you'd practice, training yourself to look out for what made better cattle, hogs, sheep, or horses.

Ben got to travel around the region and beyond, learning how to judge, but also how to express himself, how to build a better vocabulary, how to talk to adults without looking down or swallowing his words. He got to see some neat ranches and meet a lot of good people, too. He may have been from a small town, but his horizons were sure expanding. What's more, the poise he acquired through judging would serve him well decades later when he was asked to testify before a traveling U.S. Senate committee looking into range reform.

This happened about the time that all those environmental know-it-alls had taken up the battle cry of "Cattle-Free by '93," and were making noise about banishing herds from public lands because they claimed that cattle degraded it, overgrazed it, ruined the pristine streams, crapped all over it. Any real rancher knew that was just a bunch of bullshit — the environmentalist claims, not what the cows left. Any real rancher knew that, properly managed, cows made the land stronger and healthier.

And any real rancher knew that, a lot of times, it was *because* of environmentalists and a lot of other bureaucratic strangleholds that his land and livelihood were in deep trouble. Ben looked those Washington people straight in the eye and told them that. Also told them that, while everybody seemed concerned with a few vanishing animal species, it was young ranchers who were "an endangered species." That with soaring land values and crazy debt service, "in no way, shape or form can any young person start out in ranching in today's world; it's not possible. It's nothing more than a dream anymore for anybody — and that's as close as you're gonna get to it — a dream. If you're not born into it or don't inherit the land, you better forget it and go look for something else to do." And while he was at it, he informed the committee that the demise of ranchers meant "We're in danger of losing some of the best fabric that this nation has. Our value system, the things that made this country great — we're in danger of losing them."

Was he nervous talking in front of government big shots like Sen. Alan Simpson of Wyoming? Hell, no. Why would he have felt overmatched going up in front of a bunch of lawmakers in suits and ties? He'd learned what it's like to go up against big boys in helmets and shoulder pads when he was a 150-pound offensive lineman for the Meeker High School Cowboys. Learned it wasn't all about the other guy's size or reputation; your own toughness and conviction counted plenty. Look a man straight in the eye and don't back down. Keep moving forward because that's about the best direction there is.

After he graduated high school, Ben left his hometown for a while, starting out at Northeastern Junior College in Sterling. Two years later, he figured since he was going to spend the rest of his life in Meeker, why not see another part of the country? Hello, California Polytechnic Institute in San Luis Obispo. Graduated with a degree in animal science. Almost as important, while he was there he learned how to make killer barbecue, a skill that would come in handy down the road.

He returned home, diploma in hand, in the spring of 1990. It was "a kind of gloomy-doomy day," snow season going to wet season, mud everywhere. Ben walked up to his father, who was in the fields. Instead of a congratulations, the son got a "Glad to see you. Now, isn't about time for you to get in your old clothes and get busy helping?"

So he did. Of course, it wasn't all that easy now. Both men were "stubborn and hard-headed and we had our go-arounds." Ben had learned plenty at school and wanted to try new things. Allan had worked the land for forty-plus years and he wasn't so sure he wanted to try all those new things. Still, there was love and grudging respect between them and they soldiered on.

He didn't know when exactly, but one day Ben began to notice that Allan's shoulders weren't as big as they used to be; his "go-go-go" father was starting to slow down and fade. It wasn't exactly Alzheimer's, but something pretty close. Physically, Allan was holding his own, but mentally — damn, watching him in the final years, well, it just wasn't right. He was the kind of man who should have died quickly; maybe a heart attack up on his horse. Lingering like a ghost was a cruel business. But justice isn't necessarily a staple of ranching life; you hope things even up in the end, but you can't expect nature to be fair. But when the old man died in 2005 at least it was a mercy.

By the time he lost his father, Ben had already lost a wife. Shelley had come from back East originally; she wasn't born to the ranching life. That might have had something to do with it. She and Ben became parents together when Cole was born. But a son wasn't enough to keep them together as man and wife.

After the divorce, Ben was at loose ends for a while. Then he caught sight of a local girl. Hmmm. He asked around. Jamie — that was a real pretty name. She was born in Meeker, a ranch girl. Easy on the eyes, for sure, but, damn, she was a lot younger than him—about sixteen years younger. Still. . . .

It just so happened there was this Fourth of July dance in town coming up. He managed to get a turn or two with her. Pretty soon, he got around to calling, asking her if, y'know, she'd like to go out. She wasn't home, so he left a message. She called him back and left him a message. He played the tape about ten times. He had it bad.

They went out on a date. Then a lot of dates. It started getting serious. He was a little worried about that difference in their ages. Tried to joke about it, telling her she'd have to be willing to "wheel me around in my wheelchair and change my Depends." Jamie didn't back away. Less than a year after that first Fourth of July dance, they were married. They moved into Ben's house, the house his grandparents built in 1950, the house his grandmother was thrilled about because it was the first time she'd had running water. About a year after that Charlie Ann—Charlotte Anna on her birth certificate—came along.

Jamie taught kindergarten part-time. Ben worked the land. He made extra money running a barbecue business. Weddings, dances, big parties, small parties—he didn't advertise, just relied on word of mouth. Business was pretty good, but at least once a year it got *real* good. That's when he'd gather some of his scrub oak (that was the secret to San Luis Obispo barbecue) and handle the catering for a big shindig thrown by Henry Kravitz, a financier from New York. Some said he was a billionaire; owned several thousand acres in the valley. Came up for a month in the summer. Threw some kickass parties for friends and acquaintances. A lot of these people were what you could call famous. Ben Rogers, Meeker kid, served the likes of Michael Douglas, Barbara Walters, Oscar de la Renta, New York City Mayor Michael Bloomberg, and U.S. Vice President Dick Cheney. Sometimes, it would hit him because, "It's a pretty intense atmosphere; your body gets a whole 'nother feeling. It's like *whoa*."

Being around celebrities was interesting; making money from the catering was better. See, when you ranch for a living, extra income is always a good idea. That's why he rented out some of his land for a gravel pit. Hey, it was out of the way, nobody could see it, the money was pretty much guaranteed. That was no small consideration in a dicey business, where, Ben knew, "you gotta diversify, you gotta be on your toes, you gotta be aggressive, you gotta stay current with the times, and you gotta be lucky."

As much as he appreciated the extra paydays from Mr. Kravitz, Ben might have appreciated something else almost as much. See, working for Mr. Kravitz was a powerful reinforcement of a lesson that Ben had built his life around, a simple lesson he'd learned from his father and other good people. You work hard and you show people respect, and it doesn't matter who they are or how much money they have, they should do the same for you. What's more you should trust them to do that; you should be able to offer a man your hand on a deal without having to write every little last thing down in some contract. Those were ranch values. The values Ben acquired as a boy and never let go of as a man. The values he wanted Cole and Charlie Ann to grow up with. Values that would "put a good foundation under 'em."

Values that—like the life and the land that spawned them—you should honor and never take for granted.

The wind is murmuring, almost like it's providing some kind of background a capella chorus to the cadence of Ben Rogers' voice, which is involved in one of its frequent paeans to the land, the ones that come so often it's hard to keep count, the ones where he gets so emotional the words and thoughts bump into each other. Right now, he's talking about "Probably what I love best about ranching is being out here in God's creation. I mean, what a church I'm in everyday! It's so unreal. I could be out here and I see some of the most beautiful sunrises in the morning. And all through the day, whether we're gathering our cattle off the allotment and you're on horseback and just riding and looking at all the changing colors and, y'know, sometimes I'll just stop to look at a leaf; go over to a bluff somewhere and just look at the landscape; just soak in the beauty. I guess I never take this land for granted."

He's up near the summer allotment when he stops the '87 Ford truck with the cracked windshield—all his vehicles have cracked windshields—and "oh, maybe 286,000 miles" under its hood, and gets out. He sees some wild honeysuckle. "Cole really likes that; I better cut him some."

Sooner than later, he hopes, there will be other things about his land and life to tie him closer to his son. He wants to share special times. Like fall, when he and the men who help him gather the cattle sleep in a cabin up in the high country and "We cook a big meal together in the morning, and you have a nice breakfast before you leave all day, and you come back in the late afternoon and you might have a beer, y'know, just relax and enjoy it and look at the splendor of the fall. And all the work you put in with the cattle and calves—you can see it. It's early October and it's already gotten cool so their hair is starting to grow out again, and they're fat because they do so well in August and September because some of the best grass in the world is in Rio Blanco County. And you can see how much weight they've put on and, it's just like your reward for all your hard work. It's like nature is just telling you, 'Man, look what we did together, look what we did!' "

When he starts in like this, his words become more than words. You can see the fall colors—the reds and golds—in the tree-filtered sunlight. Feel the cool tang of morning on your tongue. Taste that glacier-cold beer in the afternoon. Smell the woodsmoke curling around your nostrils as everything turns dark.

Sometimes the scene fast-forwards, and his words take you to late winter and early spring, when "There is nothing like calving season, seeing that new crop of calves come. I love that, I love those baby calves coming, seeing all the work and time you put in selecting bulls and seeing what you get, seeing those calves and those cows mother up to those calves, the amazing ability they have to sense which calf is theirs when they all look alike. It's pretty cool."

Equally cool is the way the land is so willing to be your partner if you treat it well. "Yeah, the land is a living, breathing, organism," says Ben, voice getting a little softer. "Every year it rejuvenates itself. We have snow and that moisture turns the grass green in the summer and the cows eat it, but that plant has got the ability to regenerate."

He stops. Tilts his head back and nods a little, looking back in time.

"I wrote that once in an essay in junior college, about the cycle of things. Got an A, too. Only time I got an A in English."

He laughs a little, but not for long. He's looking around at so much beauty it's reminded him of something he wants to say, and soon he's off on another soliloquy of magic and wonder and the words start colliding and you hear him say, "I feel like I'm just a, well, kinda like a grain of sand in time. I'm just a small minute little entity that's come along and happens to be on this land for just a very short time and, yeah, I've got a title to it in the scheme of how our world is set up right now, but that really probably doesn't mean anything because I'm just here, well, hopefully I'm here to make it better than it was when I'm gone. That's what I wanna do. I mean, weed management, or increasing our grass production, or managing streams, or planting trees—every year I allocate time to do something to make things better because we're just gonna be here for a little while. We're just short-timers."

But still here longer than most. Look at the PIONEER license plates he has on his truck and hear him say, "I take pride in having that." Just like he takes pride in mentioning that the Russell Ranch is older than a century and "I think it says a lot in this society and today's world about some longevity and some stick-with-it ability. It shows that if you love what you do you'll find a way to make it work. There's been a lot come and go in this valley, but we're still here."

The problem with being around so long is it sometimes makes you more sensitive to people who haven't. Not that Ben demands some kind of ranching pedigree of his neighbors. There's plenty of good people around. Take Greg Norman, the famous golfer who owns a massive chunk of land near the Russell Cow Camp. Why, one time, Ben saw Norman in a sandwich place in Meeker, introduced himself as a neighbor, and found Norman to be just as natural a fellow as you could want.

But sometimes there is a problem. For instance, while Greg Norman was cordial to Ben, it's not Norman who Ben had to deal with that time when a fence got busted and some of Ben's cows wandered onto Norman's property. No, it was one of the managers of the property, and they seem to change so often it's hard to build any kind of relationship. There's also the problem that, as Ben says, "There's not very many real true ranching people left. You still have some ranches, but a lot of them are trophy ranches. They may have cattle and raise hay, but they have a manager—or several managers—running the ranch because the owner lives in New York or Houston or Sydney. And some of the people that run these places are good people to be around, and they're trying to be neighborly and do a good job, but some of them you wouldn't even want to get near their property; they're a bunch of idiots."

A lot of times the idiots managing a big ranch aren't ag folks. No, often as not it's someone associated with the owner in some roundabout way. Someone "who thinks he can just step in and be the ranch manager and you can't." There's one ranch in the valley that got sold and the person who wound up being the manager used to be the cook. And even though "He wears this big cowboy hat and pretends he's a cowboy, he doesn't have the qualifications to run a ranch, he's not a real ag person. But the owner thinks he's a great guy so he gets to run it anyway."

He shakes his head. Time to move on. Complaining isn't his way. Neither is feeling sorry for himself.

"Sometimes ranchers may look for sympathy and say, 'Oh, what a tough life we've got,'" he says. "But you know what? You don't *have* to do this. Nobody's pointing a gun at your head. It's a life we choose. Sure, it's got hardships at times, but that makes every day more interesting."

So, no, he's not about to pretend he's a day away from bankruptcy. That'd be a lie because "The value of the land has increased over the last ten or fifteen years, even the last five years." Of course, it isn't money he can exactly spend because even with those 1,280 acres and about 500 head of cattle, "One of my sayings is, 'land-rich, cash-poor.' I go by that every day. Yeah, I got lots of assets, but they're tied to the land. And if I liquidate them, I don't have my life."

If it comes to that, well, it'll be a sad day for Ben Rogers. Just like it'll be a sad day if it works out that neither Cole nor Charlie Ann decide to stay on the land. Their father will understand, of course. He knows that ranching is a life you've got to want, not tolerate. So, for now, he'll content himself with looking forward to taking his son—maybe his daughter, too—to some new place by Elk Crick where the fly fishing is just perfect. He'll look forward to maybe bringing his children to a cattle sale, listening to what they have to say about selecting bulls. He'll hope that whatever the future brings, he'll have the opportunities to share with them the splendor of the land, the exploding fall colors, the flowers that are so delicate and beautiful, the cattle that have grown fat like bad weeds. He'll hope that someday his children will be able to find enough welcomeness to comfort them. And if, one day in the future, it comes down to making a decision about whether to sell the land or protect it in an easement or do something else, well, the future is when he'll make that decision.

But right now, he's made another one.

There's this place he wants to show off, one of those spots where the oak and aspen and flowers and grasses festoon the land, a place pretty enough to make a man stop whatever he's doing and just *look*. Only problem is in order to reach it, well, you have to sorta trespass on Greg Norman's property. So Ben steers his truck off the road and does just that. He knows that Norman's security people are probably watching. He knows that "We might get in trouble for this." He knows this but he doesn't care. "Hell, we'll just tell 'em we're looking for the manager or something."

Then he grins. A boyish grin. Maybe the same grin he had when he and Patch were chasing frogs, racing ahead of his father and grandfather to get the first cold drink from a spring bubbling out of the ground.

Then he laughs. A mischievous laugh. Maybe the kind of laugh a grown man makes when he figures that sometimes in order to gain access to a beautiful church, you need to have a little bit of the devil in you.

Sunset

AUBERT RANCH

- *Mesa County*
- *Owners: Maxine Aubert, Chele Aubert Hawks and Dave Hawks*
- *First ranched in the 1880s*
- *Conservation easement held by Mesa Land Trust*

Aubert Ranch exists in two parts—each in a completely different ecosystem, yet only ten miles apart. Both parts lie in Glade Park, a large mesa just a few miles southwest of the city of Grand Junction. The lower ranch is characterized by sage meadows and spectacular eroded white sandstone formations, including several natural arches. Its drainages descend into the Colorado River Canyon to the north. Ten miles south, the mesa ascends into ponderosa pine, aspen, scrub oak, and even spruce and fir tree habitat. Here, the upper part of the Aubert Ranch extends all the way to the edge of Unaweep Canyon, where views exist of Utah's La Sal Mountains.

French Basque sheepherder Florenz Aubert immigrated to America in search of land, in short supply in the French Pyrenees, and landed in Glade Park in 1918. Many Basque came to America after World War I, settling in western Colorado to ranch for sheep. Florenz's sons August and Lawrence continued sheep ranching until 1985, when the family switched to cattle. August's daughter Chele and her husband, Dave Hawks, carry on the Aubert tradition today.

Cows, Unaweep Canyon, and La Sal Mountains at sunrise

Lawrence Aubert, 1937

Sandstone arch

Unaweep Canyon

WEST SLOPE

ESCALANTE RANCH

- *Mesa and Delta Counties*
- *Owner: Dick Miller*
- *First ranched in the 1880s*

Much like Aubert Ranch, Escalante Ranch spans two distinct ecosystems. Its lower parts span the red sandstone canyons of the lower Gunnison River, its upper portion the aspen-covered Uncompahgre Plateau. Escalante Creek connects these sections, and there is a very rugged, twenty-five-mile road used to get from the bottom to the top of the ranch. At one point along the road, Escalante Canyon is 1,300 feet deep. Desert bighorn sheep and mountain lions populate the lower canyon, elk, deer, and black bears the upper reaches.

Autumn on top of the plateau is as magnificent as anywhere in Colorado. The connectivity of the ranch allows the cattle to be summered in the meadows of the Uncompahgre Plateau and driven by cowboy down to the relatively balmy environs of the Gunnison River in winter. Historically, Ute Indians wintered in the middle part of the canyon, near what became the settlement of Escalante Forks, until being forced out in the 1880s. The Musser family then settled the canyon, and farmed and ranched it for three generations until Dick Miller purchased their ranch in 1990.

Desert bighorn sheep

Indian rock art

Escalante Forks

Cow camp on the Uncompahgre Plateau

Escalante Creek

Ranch bridge across the Gunnison River

Gunnison River

SUNRISE CANYON RANCH

- *Delta County*
- *Owners: Tom and Bonnie McCluskey*
- *First ranched in the 1880s*
- *Conservation easement held by Colorado Open Lands*

The Smith Fork of the Gunnison River drains from the West Elk Mountains southeast of the town of Paonia. Twenty miles later is meets the Gunnison in the Gunnison Gorge Wilderness. Sunrise Canyon Ranch lies along the Smith Fork, just above the wilderness in its own spectacular canyon. The fork irrigates hay meadows and wetlands in the canyon bottom, creating an isolated paradise far from human activity. The canyon rim affords views of the West Elks to the east and the Uncompahgre Plateau to the west.

Ranch headquarters looking west down Sunrise Canyon

Sunrise Canyon looking east to the West Elk Mountains

CENTENNIAL RANCH

- *Ouray County*
- *Owners: Vince and Joan Kontny*
- *First ranched in 1879*
- *Conservation easement held by Colorado Cattlemen's Agricultural Land Trust*

Centennial Ranch lies along the banks of the Uncompahgre River, twenty-five miles south of the town of Montrose. The Uncompahgre comes out the San Juan Mountains, which back up to Red Mountain Pass south of the town of Ouray. Hay meadows and stately cottonwood trees along the river define the scenery of the ranch.

Canadian James Smith first settled in Colorado's Wet Mountain Valley, then migrated to the Uncompahgre River Valley in 1879, where he homesteaded what is today Centennial Ranch. Smith and his wife, Charlotte, raised nine children on the ranch, as well as apples, peaches, and hay. They also raised horses on land they owned in the high country. Smith's descendents grazed cattle on public lands and wintered them at what was then called Smith Ranch. The legacy of the Smith family ended when Vince Kontny bought the ranch in 1992 to use as winter ground for cattle that spent their summers at his Last Dollar Ranch above the town of Ridgway.

Restored barn

Uncompahgre River

LAST DOLLAR RANCH

- *Ouray County*
- *Owner: Rod Lewis*
- *First ranched in 1888*
- *Conservation easement held by American Farmland Trust*

Last Dollar Ranch lies on Hastings Mesa, west of the town of Ridgway. It is remarkable for the mountains that rise above it, the Sneffels Range. Colorado's scenic Dallas Divide along State Highway 62, the route to the historic town of Telluride, is famed for its views of Mount Sneffels and the magnificent Double RL, a ranch owned by Polo founder Ralph Lauren. Because it lies out of sight above the highway, Last Dollar Ranch is not as well known, though it's just as spectacular. On the other hand, it is a famous place—it's just that people don't recognize it as the most frequently used location for Marlboro cigarette and Budweiser beer ads!

Sam Nesbitt homesteaded the ranch in 1888. In 1901, the Collins family acquired it and worked it for several generations. In the beginning they built ditch irrigation systems and raised hay, wheat, oats, and barley, as well as milk and range cows. It remained in the family until conservation easement pioneers Vince and Joan Kontny bought the ranch in 1989. The Kontnys performed a remarkable restoration of the old ranch buildings that make Last Dollar Ranch one of the most historic in the West. They sold the ranch to Rod Lewis in 2007.

Restored barn

Ranch headquarters

Sneffels Range

Sneffels Range

Below: *Mount Sneffels on the left*

BEAVER MESA RANCH

- *San Miguel County*
- *Owners: Howard Hughes family*
- *First ranched in the 1880s*
- *Conservation easement held by Colorado Cattlemen's Agricultural Land Trust*

Beaver Mesa Ranch lies twenty miles west of the town of Telluride and south of the San Miguel River. Located on an isolated 8,000-foot mesa, it affords some of the most magnificent views in a region known worldwide for its views. The Sneffels Range can be seen to the east, and Little Cone and Lone Cone mountains are in the ranch's backyard. Beaver Mesa Ranch is surrounded by the zigzag fences made of aspen boles that have long been ubiquitous in the San Juan Mountains, but are quickly being replaced with post and wire fences.

Dan Hughes acquired the ranch in the 1930s, first raising sheep, then cattle. In order to irrigate hay meadows, the Hughes family constructed the J & M Hughes Ditch, a monumental task given its origins twenty-two miles away at Woods Lake. Third-generation rancher Howard Hughes manages the ranch today.

Flock of crows

Little Cone (11,981')

Lone Cone (12,613')

Sneffels Range at sunset

Little Cone

SCHMID FAMILY RANCH

- *San Miguel County*
- *Owners: Schmid family*
- *First ranched in 1882*
- *Conservation easement held by San Miguel Conservation Foundation*

Iconic Wilson Peak, at 14,017 feet, is the most conspicuous, if not scenic, mountain in the Telluride, Colorado area. It is visible from every mesa and ridge above the town, and is best known for its appearance in dozens of Coors beer ads and television commercials. More importantly, it is the immediate backdrop for the Schmid Family Ranch, first worked by the family in 1882, and still worked today by descendent Marvin Schmid and his family. The Schmids got their start in the area when they followed 19th-century miners to Telluride and grew hay for what must have been thousands of horses, mules, and burros—the mining stock. The arrival of the Rio Grande Southern Railroad in 1890 ended the need for local hay—it could be shipped in! Today the Schmids raise beef cows. In a region where ranches of this significance are bought and sold for eight-figure prices, it is a testament to the Schmid family's ethical values that this place is still a ranch, not a subdevelopment.

Sneffels Range to the north

Wilson Peak (14,017')

Find the barn cat!

REDBURN FLYING R RANCH

- *Montezuma County*
- *Owner: Pat Redburn*
- *First ranched in 1891*
- *Conservation easement held by Montezuma Land Conservancy*

The Flying R is bisected by the Dolores River and lies fifteen miles northeast of the town of Dolores. The Dolores River drains from the San Miguel Mountains to the east and is lined with some of the most beautiful cottonwood trees in Colorado by the time it reaches Dolores. The Flying R lies in the part of the valley that is constrained by red sandstone cliffs, making a unique backdrop to the ranch. James Ben Millard assembled the ranch in 1908. The barn that he built that year, in part from blocks cut from the nearby sandstone and from large red spruce logs harvested from the ranch, stills stands. Millard raised cattle, sheep, and potatoes, as well as the hay he filled the barn with for fifty years. The Redburn family acquired the ranch in 1982.

Dolores River

Millard barn, built in 1908

International Standard Book Number: 978-1-56579-637-9

Designer: Mark Mulvany Graphic Design
Editor: Ali Geiser
Project Management: Charlie Clark Books, LLC

Published by:
Westcliffe Publishers,
a Big Earth Publishing company
1637 Pearl Street, Suite 201
Boulder, Colorado 80302

Printed in China by C & C Offset Printing Company

9 8 7 6 5 4 3 2 1

Library of Congress Cataloging-in-Publication Data:
CIP data on file.

Printed on Korean Moorim Neo matt art paper that comes from sustainable forests managed under the Forest Stewardship Council.

Typefaces used are various forms of ITC Cheltenham and Futura Bold.

For information about John Fielder, his fine art photography galleries, photography workshops, and public appearances, please visit **johnfielder.com.**

For more information about other fine books and calendars from Westcliffe Publishers, a Big Earth Publishing company, please contact your local bookstore, call us at 1-800-258-5830, or visit us on the Web at **bigearthpublishing.com.**

The author and publisher of this book have made every effort to ensure the accuracy and currency of its information. Nevertheless, books can require revisions. Please feel free to let us know if you find information in this book that needs to be updated, and we will be glad to correct it for the next printing. Your comments and suggestions are always welcome.

Photographic Information

The images in this book were predominantly made with two cameras—a Linhof Master Technika 4 x 5 field camera and a Canon G-9 digital point-and-shoot camera. Almost all of the large reproductions were made with the 4 x 5 using Fujichrome Velvia 100 transparency sheet film and an assortment of view camera lenses ranging in focal length from 75mm to 500mm—equivalent to 22mm to 190mm in 35mm format photography. Most of the smaller images were made with the 12.1-megapixel Canon, which has a zoom lens with optical focal lengths of 35mm to 210mm. A few of the larger reproductions were also made with the Canon, and it is remarkable that they appear no less sharp than the 4 x 5s. Most of the wildlife images in the book were made with my Canon EOS film camera with longer telephoto lenses.

For the first time in thirty-nine books, I made my own scans from the transparencies, using an Epson Perfection V750 Pro flatbed scanner. This scanner and the Canon G-9 are very affordable, and some of their features are examples of the sophisticated technology available to amateur and professional photographers alike that a short time ago existed only in the commercial domain. I used Photoshop, a digital graphics editing program, to adjust both the scans and the digital photographs in order to represent what I actually saw at the ranches. Raw scans do not look like reality—they are only a starting point to ultimately match the final file to the transparency, if not the scene itself. Inherent in both transparency and scan is a more narrow contrast range from highlights to shadows than what our eyes perceive, and it must be widened to manifest detail in those light and dark areas. Even the original digital files made by the Canon camera must be improved to achieve reality.

It is very important to understand that this remarkable tool, Photoshop, and other programs like it, can be used to manipulate and misrepresent reality. Photographers are often tempted to make things look better than they might actually have appeared in person, such as by enhancing color saturation. While I do approve of manipulating reality for artistic purposes—this freedom is inherent in art—I am not a supporter of manipulating reality in photojournalism. In part because I consider myself a realist in the artistic sense, but also because of my political advocacy, I do not manipulate my photos beyond basic color and exposure corrections. I use my photographs to try to influence people to become advocates for protecting the natural, and in this case rural, landscapes. Therefore, my credibility, and the value of what I say and write, depends upon people believing that what I represent in my photographs is real, if not truth. —**J.F.**

Colorado Coalition of Land Trusts Organizational Members

*** Accredited by the Land Trust Accreditation Commission as of March, 2009; * Applied/Registered for Accreditation in 2009*

Land Trust Members	PHONE	COUNTIES COVERED	WEBSITE
**Aspen Valley Land Trust	970-963-8440	Pitkin, Garfield, parts of Eagle Gunnison	www.avlt.org
*Black Canyon Regional Land Trust	970-252-1481	Delta, Montrose, Ouray, Gunnison	www.blackcanyonlandtrust.org
Clear Creek Land Conservancy	303-526-1151	Jefferson, Clear Creek, Gilpin	
**Colorado Cattlemen's Agricultural Land Trust	303-225-8677	Statewide	www.ccalt.org
**Colorado Open Lands	303-988-2373	Statewide	www.coloradoopenlands.org
Colorado Water Trust	720-570-2897	Statewide	www.coloradowatertrust.org
Colorado Wildlife Heritage Foundation	303-291-7212	Statewide	www.cwhf.info
The Conservation Fund	303-444-4369	Statewide	www.conservationfund.org
Continental Divide Land Trust	970-453-3875	Summit, Park	www.cdlt.org
Crested Butte Land Trust	970-349-1206	Gunnison	www.cblandtrust.org
Crestone/Baca Land Trust	303-442-8678	Custer, Saguache	www.crestonelandtrust.org
*Douglas Land Conservancy	303-688-8025	Douglas, Elbert, Jefferson	www.douglaslandconservancy.org
Ducks Unlimited, Inc.	970-481-7793	Statewide	www.ducks.org
**Eagle Valley Land Trust	970-524-0870	Eagle	www.evlt.org
**Estes Valley Land Trust	970-577-6837	Estes Park	
Gunnison Ranchlands Conservation Legacy	970-641-4386	Gunnison, Saguache	www.gunnisonlegacy.org
Land Trust of the Upper Arkansas	719-539-7700	Lake, Chaffee, Fremont	www.ltua.org
*La Plata Open Space Conservancy	970-259-3415	La Plata, Montezuma, Dolores, Archuleta, Ouray	www.lposc.org
Legacy Land Trust	970-266-1711	Larimer, Jackson, Weld	www.legacylandtrust.org
**Mesa Land Trust	970-263-5443	Mesa	www.mesalandtrust.org
*Middle Park Land Trust	970-887-1177	Grand	www.middleparklandtrust.com
*Montezuma Land Conservancy	970-565-1664	Montezuma, Dolores	www.montezumalandconservancy.org
Mountain Area Land Trust	303-679-0950	Jefferson, Park, Clear Creek	www.savetheland.org
The Nature Conservancy	720-974-7007	Statewide	www.nature.org/colorado
Orient Land Trust	719-256-5212	Saguache	www.olt.org
The Palmer Land Trust	719-632-3236	El Paso, Crowley, Huerfano, Las Animas, Park, Pueblo, Teller, Otero, Custer, Fremont	www.palmerlandtrust.org
Rio Grande Headwaters Land Trust	719-657-0800	San Luis Valley: Alamosa, Conejos, Costilla, Mineral, Rio Grande, Saguache	www.riograndelandtrust.org
Roaring Fork Conservancy	970-927-1290	Garfield, Eagle, Pitkin	www.roaringfork.org
*San Isabel Land Protection Trust	719-783-3018	Custer, Southern Fremont, Northern Huerfano, Western Pueblo	www.sanisabel.org
San Miguel Conservation Foundation	970-728-1539	San Miguel	
South Metro Land Conservancy	303-738-0348	Arapahoe	
Southern Plains Land Trust	719-526-5276	Baca	www.southernplains.org
Southwest Land Alliance	970-264-7779	Archuleta, Mineral, Hinsdale	www.southwestlandalliance.org
Trust for Land Restoration	970-626-3236	Statewide	www.restorationtrust.org
Trust for Public Land	303-837-1414	Statewide	www.tpl.org
**Wilderness Land Trust	970-963-1725	Statewide and National	www.wildernesslandtrust.org
Yampa Valley Land Trust	970-879-7240	Routt, Moffat, Jackson, Rio Blanco	www.yvlt.org

Local Government Members	PHONE	COUNTIES COVERED	WEBSITE
Aspen Parks & Recreation	970-429-2034	Pitkin	www.aspenrecreation.com
Boulder County Parks & Open Space	303-678-6200	Boulder	www.co.boulder.co.us/openspace/
City of Boulder Open Space and Mountain Parks	720-564-2038	Boulder, Jefferson	www.ci.boulder.co.us/openspace
City of Fort Collins	970-221-6600	Larimer	www.ci.fort-collins.co.us/naturalareas/
City of Loveland	970-962-9203	Loveland and Surrounding Area	www.ci.loveland.co.us/parksrec/OpenNatural.htm
Douglas County Open Space	303-660-7495	Douglas	www.co.douglas.co.us/openspace/
Jefferson County Open Space	303-271-5925	Jefferson	http://co.jefferson.co.us/openspace/
Larimer County Rural Land Use Center	970-498-7686	Larimer	www.co.larimer.co.us/rluc/
Larimer County Parks and Open Lands	970-679-4575	Larimer	www.larimer.org/parks/
Lower Arkansas Valley Water Conservancy	719-254-5115	Otero	www.lavwcd.org/
Pitkin County Open Space and Trails	970-920-5203	Pitkin	www.aspenpitkin.com/depts/21/
San Miguel County Open Space	970-728-3083	San Miguel	http://openspacerec.sanmiguelcounty.org/
Summit County Open Space & Trails	970-668-4060	Summit	www.co.summit.co.us/OpenSpace/
Town of Breckenridge	970-547-3110	Summit	www.townofbreckenridge.com/

WEST SLOPE
ROCKY MOUNTAINS
FOOTHILLS
GREAT PLAINS
VERMILLION
FETCHER
287
ROBERTS
EAGLE ROCK
SWEDLUND
Walden
Sterling
Maybell
Craig
Steamboat Springs
Ft. Collins
40
14
76
MANTLE
PARKVIEW MOUNTAIN
25
Ft. Morgan
R&R
WRAY
34
Kremmling
Meeker
RUSSELL
YUST
Wray
13
PINEY PEAK
9
71
Denver
Vail
HILLSIDE
Eagle
Dillon
70
Glenwood Springs
Rifle
COLD MOUNTAIN
DEER VALLEY
LOST MARBLES
HARVEY
Limon
Burlington
Leadville
Aspen
24
285
DAKAN
VOLK
AUBERT
Grand Junction
385
Buena Vista
Colorado Springs
BRETT GRAY
ESCALANTE
Delta
BAR OPEN AK
SUNRISE CANYON
COGAN
FOUNTAIN CREEK
TRAMPE
Gunnison
Salida
Montrose
50
IRBY
550
Pueblo
CENTENNIAL
HILL
RIVERGATE
FLYING X
RUSK
3R
LAST DOLLAR
Ouray
Lake City
Lamar
BEAVER MESA
Telluride
La Junta
149
SCHMID
WILSON
RIO OXBOW
South Fork
Monte Vista
Walsenburg
350
SAM CAPPS
BEATTY CANYON
NOTCH
REDBURN
Alamosa
CANTRELL
JE CANYON
Springfield
WEMINUCHE
Cortez
Pagosa Springs
Durango
160
RIO PINOSO
HARTONG
CROSS ARROW
Trinidad
SALAZAR
PATTERSON
84